BRAIN ARCHITECTURE

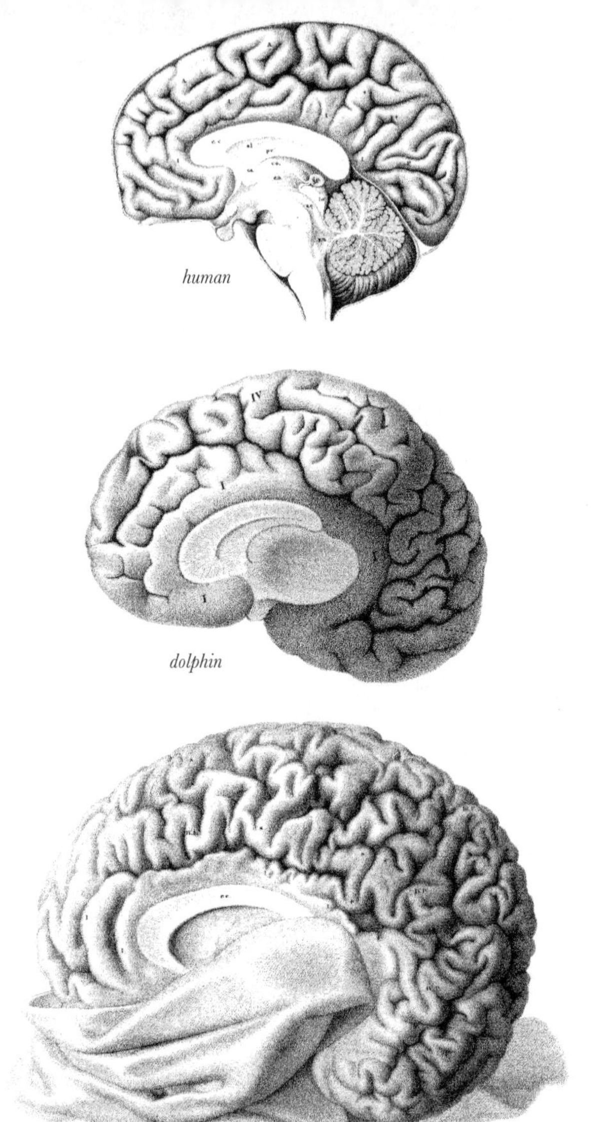

THREE MAMMALIAN BRAINS

These drawings show the appearance of the brain in three large mammals: human, dolphin, and elephant. In each case the brain has been cut into right and left halves, and a midline view of the right half is presented. In the dolphin and elephant brains, the brainstem and cerebellum have been removed for clarity. The brains are drawn to scale. From F. Leuret and P. Gratiolet. Anatomie comparée du système nerveux *(Baillière: Paris, 1857).*

BRAIN ARCHITECTURE

Understanding the Basic Plan

LARRY W. SWANSON

Milo Don and Lucille Appleman Professor of Biological Sciences
University of Southern California

OXFORD
UNIVERSITY PRESS
2003

Oxford University Press

Oxford New York
Auckland Bangkok Buenos Aires Cape Town Chennai
Dar es Salaam Delhi Hong Kong Istanbul Karachi Kolkata
Kuala Lumpur Madrid Melbourne Mexico City Mumbai Nairobi
São Paulo Shanghai Taipei Tokyo Toronto

Published by Oxford University Press, Inc.
198 Madison Avenue, New York, New York, 10016
http://www.oup-usa.org

Library of Congress Cataloging-in-Publication Data
Swanson, Larry W.
Brain architecture : understanding the basic plan / Larry W. Swanson.
p. cm.
Includes bibliographical references and index.
ISBN 0-19-510504-4 (cloth)—ISBN 0-19-510505-2 (pbk.)
1. Brain. 2. Neural circuitry. 3. Neuroanatomy. I. Title.
QP376.S86 2003
573.8′6—dc21
2002022015

9 8 7 6 5 4 3

Printed in the United States of America
on acid-free paper

From what has been said, I shall draw the only conclusion which legitimately results; namely, that the mechanism of thought is unknown to us—a conclusion with which every one will probably agree. None the less the fundamental question I have suggested exists: for what concerns us is to know whether our present ignorance on this subject is a relative ignorance which will vanish with the progress of science, or an absolute ignorance in the sense of its relating to a vital problem which must forever remain beyond the ken of physiology. For myself, I reject the latter opinion, because I deny that scientific truth can thus be divided into fractions. How, indeed, can one understand that it is permitted to the physiologist to succeed in explaining the phenomena that occur in all the organs of the body, except a part of those that occur in the brain? Such distinctions cannot exist among vital phenomena. Unquestionably they present very different degrees of complexity, but they are all alike in being either soluble or insoluble by our examination; and the brain, marvelous as those metaphysical manifestations that take place in it appear to us, cannot form an exception among the other bodily organs.

—*CLAUDE BERNARD (1873)*

To extend our understanding of neural function to the most complex human physiological and psychological activities, it is essential that we first generate a clear and accurate view of the structure of the relevant centers, and of the human brain itself, so that the basic plan—the overview—can be grasped in the blink of an eye.

—*SANTIAGO RAMÓN Y CAJAL (1909)*

NOTE TO THE READER

There are many ways to think about how the brain works, from philosophy and mathematical models at one end of the spectrum, to psychology, to biology, to chemistry and physics at the other end. Yet one thing remains common to all of them—the physical brain itself. For over 2500 years scientists have been researching the architecture, structural organization, or anatomy of the brain as an organ, the all-important organ of mind. This book is an attempt to distill the general principles that have stood the test of time, to present a new model of how the brain's functional systems are organized, and to point out how much remains to be learned about what is far and away the most complex yet intrinsically interesting object that we know of. It is written for anyone—whether computer scientist, physicist, psychologist, biologist, or general reader—interested in learning more about the basic architecture of the brain. I have taken an historical approach to give a flavor for how this problem has been approached down through the ages. It has been an exciting, heroic effort that is far from over, and it is important to appreciate that a mixture of experimental and theoretical approaches has been used from the beginning. History has shown that structure and function are simply two sides of the same coin, inexorably intertwined—both necessary and both dependent on the other.

PREFACE

No great discovery is the work of one man, or even one generation, but may represent centuries of human endeavour.

—*CHARLES SINGER (1957)*

As a new graduate student in the laboratories of the psychiatry department at Washington University (St. Louis), I was deeply impressed with the voracious eating behavior that is triggered when a specific neurotransmitter molecule like noradrenaline is microinjected directly into a tiny, very specific region of the brain—and with the equally striking drinking behavior that immediately follows the microinjection of different neurotransmitters like acetylcholine or angiotensin. What neural circuits do these chemicals activate to create sensations of hunger and thirst, appetites for food and water, and searching for these very specific goals? Later on, as a postdoctoral fellow in the anatomy and biology departments, I began to search for and analyze the hazily understood brain systems that underlie these and other types of motivated and emotional behaviors, and this interest has guided my research and thinking ever since. Twenty-five years later, at least one thing seems obvious: the explanation of any motivated behavior like eating, drinking, defending one's territory, reproducing and caring for offspring, or even sleeping—in terms of an interacting set of underlying neural systems with distinct functions—really amounts to explaining how the entire nervous system is arranged and works as a whole.

For example, a specific behavioral response like eating can be activated in many different ways (hunger, advertising, habit, and so on), goal objects need to be searched out effectively and dealt with appropriately when they are found—and then, finally, the consequences of all this activity should be used to shape adaptive ("appropriate" or "useful") behavior in the future. Experience, good or bad, counts a lot. Pleasurable things tend to be repeated, whereas unpleasant things tend to be avoided. Ever since classical antiquity in Greece, existing knowledge has been repeatedly synthesized to explain how the brain works in terms of its basic plan or blueprint, its overall organization as a system. Now is a good time to do it again because so many revolutionary insights have been gained in the last few years by a veritable army of neuroscientists—and, equally important, we are at the beginning of another profound revolution fermented by the sequencing of the human genome.

This is not a book about all the marvelous things the brain can do. That is familiar territory from personal experience. Instead, the book considers how the brain as a whole works: what its basic parts may be, and how they are related to one another as a functional system—René Descartes's wondrously sophisticated biological mechanism or machine, if you will. But in all honesty, no one at this point in time pretends to understand how the brain as a whole works. All we can do is present a model that can be tested through further observation and experimentation. The main advantage is a tangible framework for discussion, comparison, and improvement.

Today's neuroscience rests on the shoulders of three great pioneers—the histologist Santiago Ramón y Cajal (1852–1934) and the physiologists Charles Sherrington (1857–1952) and Ivan Pavlov (1849–1936). They stand out from other leaders a century ago because their prodigious laboratory work established fundamentally new principles with exceptionally broad implications for how the nervous system is organized and works. The Spaniard Cajal taught us how to describe the nervous system in terms of neural networks made up of individual units—nerve cells, or neurons—and gave us a rule based on structure for predicting the direction of information flow through them (toward axon terminals or synapses). Based on this structural foundation, the Englishman Sherrington worked out the hierarchically arranged functional organization of innate or genetically programmed sensory-motor arcs that control reflex behavior. And the Russian Pavlov described quantitatively how we form learned or conditioned—as opposed to purely reflex—responses by forming new associations between various stimuli.

Today, we can see that the second half of the twentieth century was the golden age of cellular neuroscience. The biophysics and chemistry of information processing for the individual nerve cell can be explained in great detail, especially when compared with the vague outlines of what the primary functional systems in the brain might be. As a result, many excellent introductions to cellular neuroscience are readily available, whereas overviews of systems neuroscience, especially from the perspective of fundamental brain architecture, are harder to come by. In the new century there is every reason to think that both systems neuroscience and molecular neuroscience will catch up with cellular neuroscience. The best effect this book could have would be to stimulate the formulation of alternative global descriptions of nervous system organization—which could only serve to stimulate further discussion and laboratory work. The value of a good model can be measured not only in terms of understanding basic mechanisms but also in terms of developing new cures for medical problems and of using biological principles for designing new technologies. The history of how models, hypotheses, and theories influence any branch of science is a fascinating topic, and we will return to it often in the book.

Los Angeles L.W.S.

CONTENTS

1

How the Brain Works

Structure and Function

From what has been said, the services of the brain are evident. They are of one sort according to Aristotle and of another according to Galen and his followers: look them up. It suffers ills of all kinds. Its injury is fatal, not always but most of the time.

—BERENGARIO DA CARPI (1523)

You can see these convolutions of the animal's brain [cortex] when you are at breakfast or dinner, but as to their functions both physicians and philosophers are greatly exercised. They dispute whether men have understanding through them or not.

—ANDREAS VESALIUS (1543)

There are two ways only of coming to know a machine: one is that the master who made it should show us its artifice; the other is to dismantle it and examine its most minute parts separately and as a combined unit. Those are the valid methods of learning the contrivance of a machine. . . . But, since the brain is a machine [Descartes, 1664], we need not hope to discover its artifice by methods other than those that are used to find such for other machines. There remains to be done, therefore, only what would be done for all other machines. I mean the dismantling of all its components, piece by piece, and consideration of what they can do separately and as a whole.

—NICOLAUS STENO (1669)

Most of us don't think much about our brain—let alone about how it works—until something goes wrong with it. Then we wonder why this or that distressing symptom happened and whether we can do anything about it. Stroke, depression, retardation, epilepsy, dementia, addiction, schizophrenia—the list of heart-wrenching afflictions is long indeed and doesn't even include a host of other less severe, yet frustratingly real, problems like anxiety, learning and memory disorders, attention deficits, and on and on. For answers we often turn to medicine—to neurologists and psychiatrists who usually prescribe drugs or surgery that may relieve symptoms in a more or less effective way, at least for a while. But ask 10 of the world's leading neuroscientists how the brain works—how it thinks, feels, perceives, and acts as a unified whole—and you will get 10 very different answers, unless they are very narrowly framed around the biophysics and chemistry of nerve impulse conduction and synaptic transmission. Synapses are the functional contacts between nerve cells that may change their strength based on experience. You have about 100 trillion of them in your brain, and they are so tiny they can only be analyzed with an electron microscope!

So, when it comes to explaining general principles of brain anatomy and function, there is no mystery about the uncertain state of affairs. Gram for gram, the brain is far and away the most complex object we know of in the universe, and we simply haven't figured out its basic plan yet—despite its supreme importance and a great deal of effort. There is nothing equivalent to the periodic table of the elements, relativity, or the theory of evolution for organizing and explaining a large (but still woefully incomplete and often contradictory) body of information about brain structure and function. No Mendeleyev, Einstein, or Darwin has succeeded in grasping and articulating the general principles of its architecture; no one has presented a coherent theory or model of its functional organization.

As a matter of fact, there is not even a list of basic parts that neuroscientists agree on, let alone a simple and clear account of what each part does, so how can there be a scheme for the way they are interconnected to generate our thoughts, feelings, and actions? In view of this ignorance, it is little wonder that no real cures for any of the brain afflictions mentioned at the beginning have been stumbled upon. New guiding principles based on understanding rather than fortuitous accident, and on a great deal more knowledge derived from research, are obviously needed before such cures will be discovered. This is the challenge of today's neuroscience.

We can only assume that these new principles will emerge in one way or another out of, or maybe in reaction to, what we already know about the brain, which is actually quite a lot. Many of the best thinkers throughout the long history of biology have contributed to the way we now view the physical underpinnings of behavior. By tracing the development of their basic ideas we can do two things. First, we can take stock of where we stand today; second, we can point out domains of which we are still profoundly ignorant—or could even be profoundly wrong.

If biology has taught us anything, it is that we can only understand the structure and function of the brain by considering them within the larger context of the structure and function of the body as a whole. On one hand, there is no doubt that the brain controls the body. But on the other hand, the structure of the body as it develops in the embryo, and as it evolves over geological time, profoundly sculpts the structure of the codeveloping and coevolving brain. Thus, the basic organization of the nervous system as a whole reflects the basic organization of the body. The intellectual threads of this lesson can be traced easily and directly back almost 2500 years to Aristotle—son of the physician to the king of Macedonia, student of Plato, and father of comparative, developmental, and theoretical biology—the first curator of the animal world, as F.J. Cole phrased it so nicely.

THREE BIOLOGICAL PERSPECTIVES

Aristotle's *Historia Animalium* (*History of Animals*) was the first—and, some would argue, still the finest—textbook of animal biology ever written in terms of sheer originality, breadth, logical force, and lasting influence. Aristotle seems to have written it, along with two other brilliant books (*Parts of Animals* and *Generation of Animals*), while he was in his 40s. This would have been before his final return to Athens and thus before most of his "philosophical" work was produced at the Lyceum during the last decade of his life.

Aristotle's conclusions were based on an encyclopedic treatment of personal observations on a broad range of around 500 different types of animals. Nothing even remotely approaching this scope had ever been attempted before, and it initiated observation, rather than speculation and folklore, as a basic foundation for research. And these comparative observations were not limited to just structure and function. In a strikingly modern way, Aristotle paid equal attention to patterns of behavior, to ecological interactions, and to geographical distributions.

While this comparative approach to the natural history of animals in and of itself was a major contribution to biology, it was only a starting point for Aristotle, who thought deeply about what the observations implied in terms of basic generalizations or principles. What emerged came eventually to be known as *theoretical biology*, an approach that is gaining momentum today, this time propelled by molecular genetics. Clearly, the essence of Aristotle's theoretical work has survived into modern times, but stripped by Darwin of its preoccupation with "final causes" or preordained reasons. Experience has since amply demonstrated that this teleological approach of philosophers is a very easy and very unproductive way of thinking for scientists to fall into.

In his most brilliant theoretical synthesis, Aristotle suggested that just a small number of basic body plans can account for all the vast diversity of animal forms observed in nature. Furthermore, he suggested that these body plans can be arranged in a hierarchy based on decreasing levels of complexity (with humans at the top, of course). Naturally, Aristotle's classification scheme has turned out to be incomplete and inaccurate, but, its basic form is obvious in the taxonomy graphs of any modern biology textbook. Here basic body plans are represented by the different phyla of the animal kingdom (vertebrates and assorted invertebrates: mollusks, arthropods, various worms, sponges, and so on), and the hierarchy is represented by an evolutionary tree rather than Aristotle's linear scale of life.

Aristotle's grand biological synthesis actually served as a remarkable foundation for more technical insights that are worth mentioning because they are still regarded as valid today. For example, in proposing the first scientific classification schemes for animals, Aristotle formulated the principle that judgments about affinities or similarities between animals have to be based on comparisons of all characters, not simply this or that feature. In doing this he had at least a vague notion of what we regard today as species and genera (Latin translations of terms that he actually used). Aristotle's basic concept that each major group of animals shares a common structural (body) plan or architecture extended to all the principal organ systems and included the obvious positive and negative correlations between organ systems; Aristotle's first great successor two millennia later, George Cuvier (1769–1832), based his own work on this bedrock concept. As part of this concept, Aristotle emphasized that all animals belonging to a particular class have the same basic parts, which differ only by degree; in other words, they simply may be larger or smaller, softer or harder, and so on.

There is a flip side to the unity of plan principle—the tremendous biological diversity one finds in nature. In dealing with this problem, Aristotle realized the importance of analyzing how comparisons of parts within (as well as between) major groups are actually made and interpreted. He recognized and began to articulate a fundamental difference between what we now call *homologous parts*, such as bones and teeth that have a common origin within a group, and *analogous parts*, such as human hands and crab claws that have only a superficial resemblance between major groups. The real significance of homologous and analogous parts had to await the embryologic and evolutionary work of the nineteenth century. In any event, Aristotle's abiding interest was in organs or functioning parts, rather than in the mere spatial relationships between parts. In one of his most brilliant excursions into "theoretical" biology he compared cephalopod

READINGS FOR CHAPTER 1

Aristotle, *Historia Animalium*. For English translation see: J.A. Smith and W.D. Ross (eds.) *The Works of Aristotle*: Vol. 4, *Historia Animalium*, translated by D.W. Thompson. Oxford University Press: London, 1910.

Cole, F.J. *A History of Comparative Anatomy: From Aristotle to the Eighteenth Century*. Macmillan: London, 1944. A fascinating, authoritative story is told.

Hall, B.K. (ed.) *Homology: The Hierarchical Basis of Comparative Biology*. Academic Press: San Diego, 1994. Probably more than you wanted to know, but a deep, fundamental problem that is still subject to lively debate.

Kety, S.S. A biologist examines the mind and behavior. *Science* 132:1861–1870, 1960. A brilliant contemporary discussion of possible relations between mind and biology; the story at the end, "The True Nature of a Book," is profound, funny, and classic.

Longrigg, J. *Greek Rational Medicine: Philosophy and Medicine from Alcmaeon to the Alexandrians*. Routledge: London, 1993. Here is an excellent introduction to the origins of Western medicine and biology.

Needham, J. *A History of Embryology*. Second edition, revised with the assistance of Arthur Hughes. Cambridge University Press: Cambridge, 1959. A sweeping overview of major concepts and players is presented.

Purves, D. *Body and Brain: A Trophic Theory of Neural Connections*. Harvard University Press: Cambridge, 1988. The profound influence of body structure on nervous system structure is illuminated; how else could it be?

Russell, E.S. *Form and Function: A Contribution to the History of Animal Morphology*. Murray: London, 1916. This is a truly brilliant analysis of theoretical morphology through recorded history.

Singer, C. *A Short History of Anatomy from the Greeks to Harvey*. Second edition. Dover: New York, 1957. Here is another indispensable guide to the history of structural biology.

Steno, N. *Lecture on the Anatomy of the Brain*, Introduction by G. Scherz. A. Busck: Copenhagen, 1965. A facsimile of the French edition of 1669 is presented here, along with modern English and German translations. It is my favorite essay on brain research.

2

The Simplest Nervous Systems

Neurons, Nerve Nets, and Behavior

All the organs of an animal form a single system, the parts of which hang together, and act and react upon one another; and no modifications can appear in one part without bringing about corresponding modifications in all the rest.

—GEORGE CUVIER (1789)

It is natural to imagine that over the course of evolution here on earth, which spans around 5 billion years, the simplest organisms appeared first and that, when viewed over a very long time frame, they were followed by more and more complex organisms, culminating with the appearance of modern humans only 100 thousand years ago or so (Fig. 2.1). If we continue to think along these lines, it's a good bet that we should be able to learn a great deal about the architecture of the nervous system by starting to examine it in the evolutionarily oldest, simplest organisms, then going on to analyze its organization in progressively more recent and complex organisms, before finally considering it in the ultimate puzzle, the human brain. By this time, we should have a vocabulary and set of rules that apply to all mammals, to all vertebrates, and perhaps even to all animals with a nervous system.

Unfortunately, this line of reasoning has flaws, not the least of which is that soft tissues like those associated with the nervous system leave no fossil record behind. We can detect no trace of what

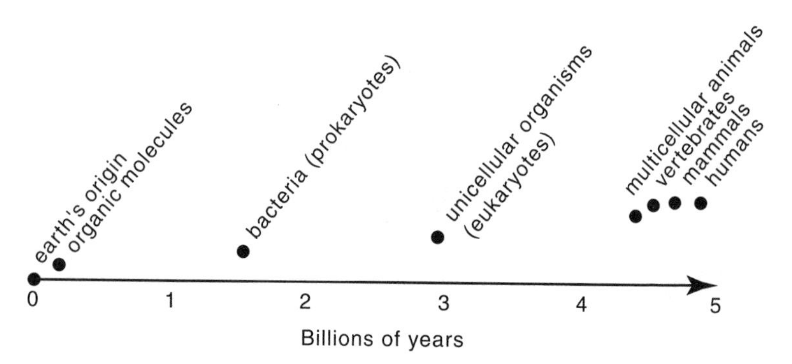

FIGURE 2.1 *The evolution of life on earth has occurred over roughly 5 billion years.*

the nervous system actually looked like in fossils of the earliest multicellular animals (called metazoans), which finally appeared on earth sometime around 600 million years ago, and it is hard to imagine how such evidence could ever be obtained. Facing up to this reality, but steadfastly remaining intrigued with the possibility of learning something basic about nervous system organization, comparative biologists have turned to modern descendants of the various early phyla of animals in an attempt to reconstruct a plausible evolutionary scenario or tree—while freely admitting that these living groups of animals have undoubtedly evolved or changed to some extent (sometimes even becoming simpler) during the long period of time since they first appeared on earth.

The most original, succinct, and appealing reconstruction along these lines was penned by the Harvard zoologist G.H. Parker in a delightful little book, *The Elementary Nervous System*, which was published in 1919, well after Darwinism had conquered all aspects of biology. The following discussion is based loosely on his general line of reasoning, fleshed out with discoveries made since then. Parker's approach was to examine each fundamental building block of the nervous system (in other words, each cell type) from the viewpoint of its basic structure and what it adds to the functional capabilities—in particular, the behavior—of the animal as a whole.

Parker's conclusion, like Santiago Ramón y Cajal's before him, was that neuronal cell types are defined by their connections within neural circuits or networks—in other words, by their inputs and (even more important) their outputs.

UNICELLULAR ORGANISMS *Behaviors Essential for Survival*

It is important to stop and realize before going on that individual species of the single-celled organisms called protozoa may be surprisingly differentiated and show rather complex behaviors—obviously, in the complete absence of a nervous system (Fig. 2.2). Fortunately, modern biology can now explain most of this behavior in terms of biochemical reactions and the molecular architecture of individual cells. It is important to know that, as is true for all cells in all living organisms, a plasma or cell membrane forms the boundary between the inside of the protozoan and its environment, and there is an electrical potential (whose maintenance requires energy from a molecule called ATP) across the membrane. Under normal resting conditions, the inside of the cell is negatively charged relative to the

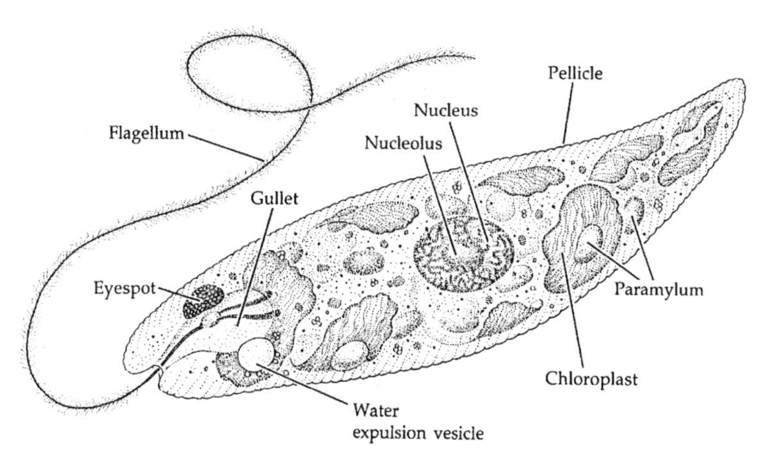

FIGURE 2.2 *The basic structural organization of a single-celled protozoan, Euglena, is shown in this diagram. Reproduced with permission from R.C. Brusca and G.J. Brusca,* Invertebrates *(Sinauer: Sunderland, 1990, p. 132).*

outside. We will return shortly to the importance of this membrane electrical polarization for information signaling in the nerve cells of multicellular animals.

Ethology is the scientific analysis of behavior, and a vast literature of brilliant research has shown that all protozoa and metazoa alike display spontaneous or intrinsic activity and that this activity includes at least three fundamental classes of behavior that are necessary for survival of the individual and the species as a whole: ingestive, defensive, and reproductive. Generally speaking, ingestive behavior is concerned with regulating internal water and nutrient supplies. In multicellular animals we usually think of these activities in terms of drinking and eating, but in single-celled animals they involve regulating the movement of water and of nutrients and waste materials across the plasma membrane and through the interior of the cell.

At the risk of stating the obvious, water is the single most important component of any cell. All biochemical reactions take place in the cell's aqueous medium, which forms at least 90% to 95% of the cell mass, and water-balance regulation is critical to the cell because there are osmotic forces (which, unregulated, cause shrinking or swelling) across its membrane, essentially because the concentration of various molecules is different inside and outside the cell.

A more or less continuous supply of nutrients to fuel metabolism is also obviously required. Metabolism generates the energy supplies for maintaining membrane potentials, synthesizing organic molecules, and so on. In protozoa, nutrients enter from the environment or are derived from intracellular organelles called *chloroplasts* that are driven by the energy from light.

Mechanisms for ingestive behavior in protozoa can be surprisingly sophisticated. For example, *Paramecia* have a channel in the plasma membrane called a *gullet* or *oral groove* that participates in taking in food, they have systems of vacuoles (membrane-bound bags) that shuttle food around and digest it inside the cell, and they even have a relatively fixed site for the expulsion of waste product vacuoles called an *anal pore* or *cytophyge.* And don't forget, protozoa can use cilia or flagella to swim toward sources of food or away from environmental threats.

The route of this locomotor activity, which we call *foraging or exploratory behavior* in "real" animals, is directed by nutrient-associated chemicals in the environment that are detected by plasma membrane receptors. These receptors are proteins that help activate the cilia or flagella in such a way as to cause the protozoan to swim toward the highest concentration of nutrient—in essence, to approach the food. Generally speaking, this type of protozoan behavior is referred to as a *taxis*, and the specific type just described is referred to as a *positive chemotaxis* because the cell swims toward the highest concentration of a particular chemical (behavior that depends, of course, on expressing the right kind of receptors in the plasma membrane).

Defensive behavior in protozoa often involves swimming away from rather than toward toxic chemicals in the environment, using mechanisms that involve negative chemotaxis. This class of behavior is obviously critical for survival of the individual protozoan and may also be triggered by other stimuli such as touch or temperature gradients. Some protozoa, especially those with a flagellum, even have a light-sensitive eyespot or stigma (an aggregation of light-sensitive pigment) that helps regulate the direction of swimming behavior (Fig. 2.2). These eyespots illustrate the principle that receptors specific for particular types of stimuli (light, touch or deformation, temperature, and chemicals, for example) can be highly localized on or in some particular region of a protozoan. And very importantly, these receptors may show adaptation—that is, the response may decrease (or perhaps increase) on repeated exposure to a particular stimulus. This leads to modified or learned behavioral responses—in other words, behavior that is altered by past experience—although in protozoa this learning does not appear to be associative (see Chapter 10).

Finally, most protozoa display both sexual and asexual reproductive behaviors. The former have the great advantage of producing in the offspring genetic variation, which is the fodder for Darwinian evolution based on a natural selection of the most adaptive individuals in a diverse population—individuals that will have the greatest probability of reproducing to perpetuate the species.

The survival of an individual requires appetitive or consummatory behaviors, as well as defensive or avoidance behaviors, whereas survival of the species requires reproductive behaviors. In humans we refer to ingestive, defensive, and reproductive classes of motivated behavior, and the search for systems in the brain that control them is still under way.

ANIMALS WITHOUT NEURONS *Independent Effectors*

Sponges are the simplest multicellular animals, and yet they took around a billion years to evolve from protozoa, arriving on the aquatic scene a half a billion years ago or so. Perhaps not so surprisingly, their body plan is more like a colony of specialized protozoa than is true for the rest of the animal kingdom, which has much more highly structured embryos with a basic architecture consisting of three stacked layers. As we shall see in the next section, these layers go on to produce distinct tissues in the adult, including nervous tissue. Sponges are so primitive and unusual that it was not until the second half of the eighteenth century that they were even recognized as animals instead of plants.

Even the simplest multicellular animals such as sponges have two considerable advantages over unicellular animals. First, their larger size provides greater resistance to physical stresses in the environment. Second, they are not, in fact, a simple colony of protozoa; instead, they have evolved different cell types—a division of labor that increases efficiency for specific tasks such as nutrition and defense.

The behavior of sponges can be described rather easily and succinctly: they are sessile suspension-feeders (Fig. 2.3). In other words, they make relatively boring pets. They are immobile, attached at their base to the bottom of some marine or freshwater environment or other, where they exchange nutrients (along with gasses and wastes) from water circulating through their body. Conceptually, their body is just an immobile, asymmetric or radially symmetric bag

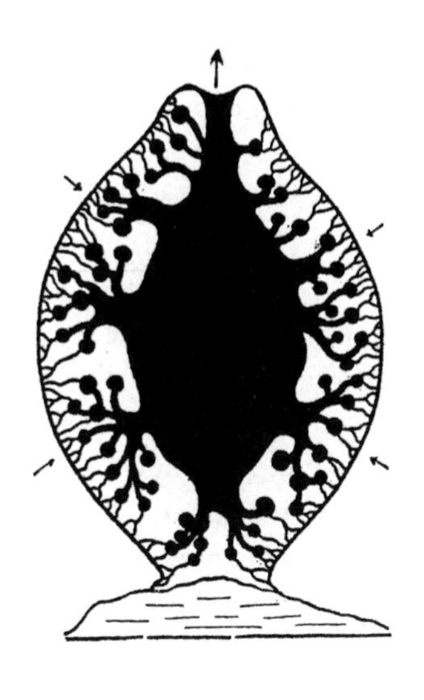

FIGURE 2.3 *The body plan of sponges is quite simple. As indicated by arrows, water flows through body wall pores into a central cavity (black) and then, through a hole in the top of the central cavity, back out into the external environment. From G.H. Parker,* The Elementary Nervous System *(Lippincott: Philadelphia, 1919).*

with many tiny holes or pores scattered throughout a relatively thin body wall. Environmental water flows through these pores into the animal's inner cavity (or spongocoel) and then out into the environment through a large hole at the top (an osculum). This circulation of water through the pores to the spongocoel and then back out through the osculum is promoted by the beating of flagellated cells that line the inside of the spongocoel.

The regulation of water flow through the sponge's body amounts to the relatively simple regulation of their feeding behavior. This is accomplished by a specialized cell type that is fundamental to our story of how metazoans control their behavior. These cells are called *myocytes*, and they have a critically important property, contractility, which allows them to shorten and thus do work. For example, these elongated cells are arranged concentrically around channels in the sponge body wall, where their contraction allows them to act as sphincters, controlling the rate of nutrient-saturated water flowing through the animal.

Sponge myocytes are probably distant ancestors of the smooth muscle cells that coat and regulate flow through our own blood vessels. For myocytes in sponges to contract and slow the flow of water, they must be stimulated directly. For example, they may contract when directly stretched, or they may contract or relax when certain chemicals interact with certain corresponding classes of receptors in their plasma membrane.

Based on functional considerations, Parker referred to myocytes as *independent effectors.* That is, myocytes (or independent effectors in general) are cells that produce a motor response when directly stimulated—without, to anticipate our story, the intervention of neurons.

Sponges are uniquely simple multicellular animals without a nervous system. However, their feeding behavior is regulated by independent effectors—smooth muscle cells whose contraction is directly stimulated by mechanical, chemical, or thermal factors. The response of these myocytes to stimuli is relatively slow and sustained, compared to the response of neurons, as we are about to see. In addition, myocytes are much less sensitive to stimuli than are neurons so that it takes a much larger stimulus to produce a response in myocytes than in neurons, and typically these responses are much slower and last much longer in myocytes.

THE FIRST NERVOUS SYSTEM *Hydra's Body and Behavior*

Jellyfish, corals, sea anemones, and hydra are among the simplest animals with a nervous system, and because of this their nervous system is the simplest to understand, as far as architectural principles are concerned. Their phylum, the Cnidaria, has a radially symmetrical body plan, and like all other animals except the sponges they have a three-layered embryo. The outer layer ("skin") faces and interacts directly with the external environment and is called the *ectoderm.* The inner layer lines a cavity inside the animal ("gut lining") and is called the *endoderm.* And a multifunctional middle layer in between, which is a primitive mesoglea in Cnidaria, is called the *mesoderm.*

For two main reasons, the common laboratory hydra provides a favorite example of how the various elements or cell types of the nervous system may have originally evolved. First, it has an elegantly simple body architecture: fundamentally, it has a mouth at one end of its body tube and a foot at the other. Second, compared to sponges, it has quite intriguing patterns of feeding and locomotor behaviors (Fig. 2.4). These animals bring food to their mouth with a set of ten-

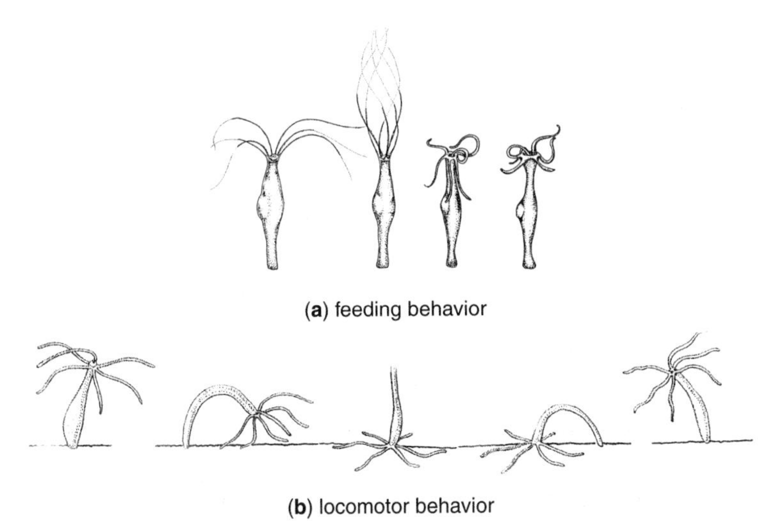

FIGURE 2.4 *Two fundamental classes of behavior in hydra, among the simplest animals with a nervous system, are illustrated here. Example (a), feeding behavior, is reproduced with permission of The Rockefeller Press from H.N. Lenhoff, Activation of the feeding reflex in* Hydra littoralis, The Journal of General Physiology, *1961, vol. 45, p. 333. Example (b), locomotor behavior, is reproduced with permission from J.L. Gould and W.T. Keaton, Biological Science, sixth edition (Norton: New York, 1966, p. 1068).*

tacles, which they also use in an ingenious way to locomote through the environment by tumbling. These complex, stereotyped behaviors require waves of patterned activity to pass up and down the body and tentacles, in coordinated ways that are not used at all by sponges, which as we have just discussed have independent effectors but no nervous system.

SENSORY NEURONS *Functional Polarity of Dendrites and Axon*

Hydra use their tentacles like paddles to wave food into their mouth, and this paddling may be initiated by food detectors near the ends of the tentacles that trigger their rhythmical movements. These detectors are sensory neurons, and they are the first of three fundamental types of neuron.

The prototypical sensory neuron is derived from the outer or ectodermal layer of the animal. It is a bipolar cell, with a detector, sensory, or input end directed toward or into the environment, and an effector, motor, or output end going to a group of responsive cells—to myocytes, for example (Fig. 2.5). The independent effectors we discussed for sponges—which have low sensitivity, slow activation, and prolonged activity when stimulated directly—can be regulated by neurons that are highly sensitive and fast acting. This means that effectors such as myocytes can now be regulated in two ways: independently as before and by the nervous system as well. In addition to speed and sensitivity, sensory neurons have the advantage that they can be highly localized in various parts of the body, like at the ends of tentacles. This provides a restricted and specialized source of inputs to effector cells or, as we shall see, to other types of neuron.

Sensory neurons beautifully illustrate two principles that are at the heart of nervous system analysis. The generality of these principles was convincingly demonstrated around the end of the nineteenth century by the crown jewel of Spanish science, Santiago Ramón y Cajal, who used a histological method developed by the Italian master, Camillo Golgi in 1873. The first is the *neuron law* (or doctrine), which is nothing more than Matthias Schleiden and Theodor

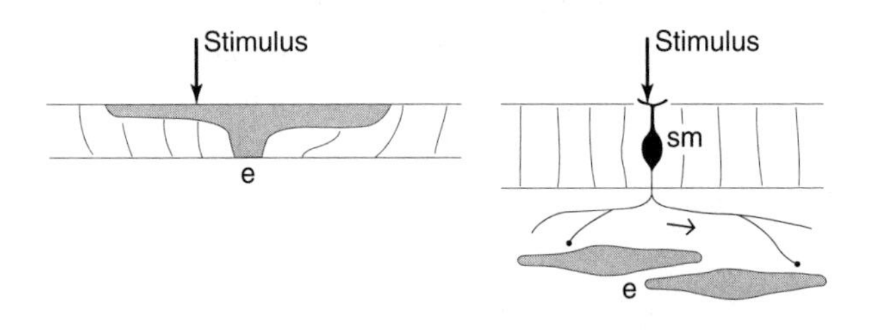

FIGURE 2.5 *Two basic cell types involved in producing behavior. (a) Independent effectors do not require activation by neurons. (b) However, independent effectors can also be activated by neurons. In this simple example, a sensorimotor neuron receives the stimulus* (arrow). Key: *Arrows show direction of information flow; e, independent effector; sm, sensorimotor neuron.*

Schwann's cell theory of 1839 applied to the nervous system. It simply points out that the nervous system is formed by a network of self-contained units or cells (neurons) that interact by way of contiguity rather than by a continuous neural syncytium, as had been thought since antiquity. The second principle is called *functional polarity* (or dynamic polarization). It states that the output side of the neuron is a single process called an *axon*, whereas the input side of the neuron consists of a *cell body* (with its nucleus and DNA blueprint for the cell) and one or more extensions of the cell body called *dendrites*. These principles allow one to model information flow through neural circuits based on the shape of their constituent neurons (axons and dendrites are usually easy to tell apart because they have distinct morphologies).

Information flow along individual neurons is by way of electrical signals conducted via the plasma membrane of the dendrites, cell body, and axon. The amplitude of these signals may be proportional to the strength of a stimulus (that is, graded potentials), as they often are in dendrites; or the amplitude may be of uniform size (that is, all-or-none action potentials—the nerve impulse), as they are in axons and sometimes in dendrites. However, it is critical to know that this information is transferred to another cell (such as a muscle cell or another neuron) via the release of chemical neurotransmitters from specialized regions of the axon called *synapses*. Mixtures of receptors in the membrane of postsynaptic cells detect the released batch of neurotransmitter molecules and go on to trigger (*a*) an electrical signal in those cells and (*b*) metabolic or molecular changes in them.

This combined/sequential electrical then chemical transmission of information is common to all nervous systems from hydra to humans. In fact, basic cellular neurophysiology is similar from hydra to humans. What changes dramatically through the course of evolution is the arrangement of the three fundamental neuron types discussed in this chapter into more and more highly organized systems or networks.

The axon of sensory neurons displays one especially important structural feature that probably applies to all axons: *divergence*. That is, every axon generates multiple synapses, usually from distinct branches or collaterals (which were first described adequately by Golgi in 1873). Because of this, individual sensory neurons can *innervate* (provide synapses to) more than one effector cell—for example, it might contract a group of myocytes rather than a single myocyte. Different branches might even innervate more than one cell type: for example, myocytes and secretory (gland) cells. Independent effectors act alone, whereas input from a sensory neuron can activate groups of "independent" effectors more or less simultaneously. In contrast, branches from more than one sensory neuron may end on one effector. This is a feature referred to as *convergence* in neural systems. One sensory neuron may innervate more than one effector cell, and each effector cell may be innervated by more than one sensory neuron.

Thus, the relatively insensitive, slow, long-lasting, and individualized responses of independent effectors may be augmented by sensory neurons, which can display extreme sensitivity, can respond quickly and rapidly, and can influence groups of effector cells. These are major adaptive advantages provided by neurons: sensitivity, speed, amplification, and coordination, not to mention the potential for highly localized distribution patterns within the animal.

MOTONEURONS *Another Distinct Neuronal Type*

In considering sensory neurons thus far, we have been describing a "one-layered" nervous system: a layer of what are usually called sensory neurons in the ectoderm or outer layer facing the environment that project or send their axons to a "layer" of effector cells like myocytes. In fact, to be completely accurate and consistent, the sensory neurons illustrated in figure 2.5 are properly called *sensorimotor neurons* because they detect environmental stimuli and project directly to effector cells. In reality, all Cnidaria actually have a "two-layered" nervous system at the very least. In the simple example illustrated in Figure 2.6, effector cells like myocytes are innervated by a second fundamental type of neuron,

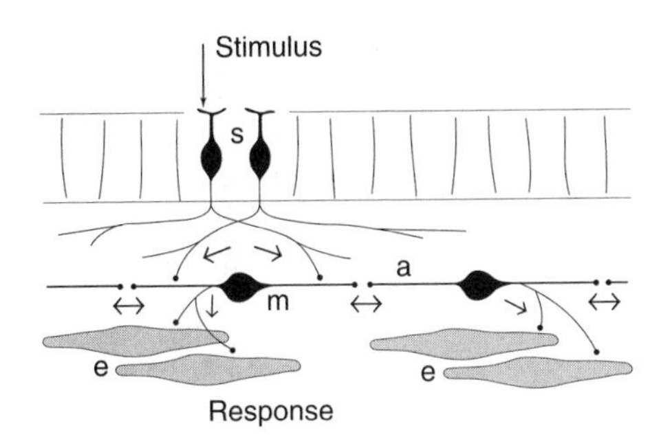

FIGURE 2.6 *The principle of a "two-layered" nervous system is illustrated here.* Key: *Arrows show direction of information flow; a, tangential amacrine process; e, independent effector; m, motoneuron; s, sensory neuron.*

motor neurons (or *motoneurons*), instead of directly by true sensory neurons, which innervate the motor neurons instead.

What are the adaptive advantages of adding a second cell type to the hydra nervous system? First, separating sensory and motor functions—a "division of labor"—adds the important possibility of regulating these two basic cell types independently. In principle, the potential for more regulation provides the potential for more complex behavior. Second, this two-stage or two-layered nervous system has even more divergence and convergence than a (theoretical) one-stage system because the axons of both sensory and motor neurons typically have multiple branches to multiple postsynaptic target cells. The situation where one sensory neuron innervates and excites multiple motor neurons, each of which, in turn, innervates and excites multiple effector cells in a "pyramid of excitation," was referred to as *avalanche conduction* by Cajal. And third, most (if not all) motoneurons in hydra interact with other motoneurons by way of tangential or "horizontal" processes (Fig. 2.6), whereas sensory neurons tend not to interact directly with one another.

Clearly, sensory and motor neurons have distinct structures and functions. Sensory neurons detect various types of environmental stimuli—chemicals, temperature, light, touch—and project to motoneurons, whereas motoneurons project to non-neural effector cells and to other motoneurons and receive inputs from both sensory and motoneurons.

This leads to a fundamental conclusion about how to classify, or really how to identify, different neuronal cell types—like classifying various species of trees or varieties of dogs. After unparalleled expe-

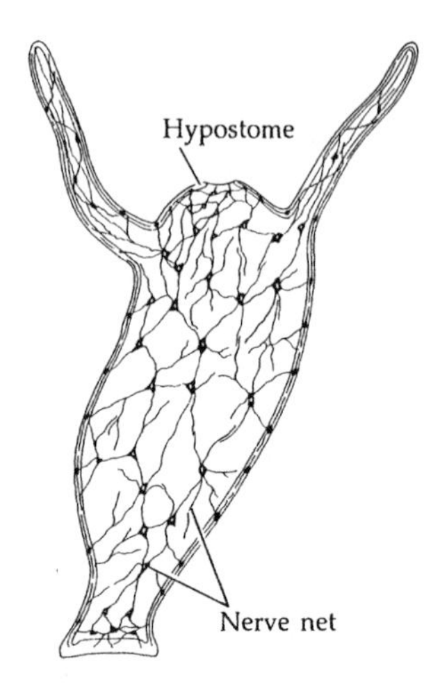

FIGURE 2.7 *The nerve net of hydra is shown here in a highly schematic way. Adapted with permission from C.F.W. Claus, K. Grobben, and A. Kühn,* Lehrbuch der Zoology *(Springer: Berlin, 1932, p. 221).*

rience and thought, Cajal came to the conclusion that the best criterion is the connections of the neuron, and especially the distribution of the axon—which in the end defines a great deal about the functional significance of a neuron. What does it innervate, and thus influence? What does it do functionally? A comparison of the basic connections of sensory and motor neurons in the hydra nerve net illustrates this principle clearly, and it can be extended to the entire nervous system. Cerebellar cortex organization in vertebrates is another terrific example; half a dozen or so very clearly defined cell types are arranged in a highly stereotyped way.

NERVE NETS *Amacrine Processes and Activity Patterns*

In hydra, motoneurons are scattered more or less uniformly throughout the body and tentacles, and they interact with one another by way of tangentially oriented processes that tend to conduct graded electrical potentials—that is, electrical potentials that get weaker the farther they spread along a neuronal process. As a result, when the nervous system of hydra is viewed as a whole, it has the appearance of what has been called for a century or so a *distributed*, or *diffuse*, *nerve net* (Fig. 2.7).

Curiously, most of the processes of hydra motoneurons fall into one of two classes: an axon that conducts electrical impulses to nonneural effector cells, and the tangential processes that conduct electrical activity between the motoneurons themselves. Under normal conditions, many of these processes conduct in either direction and have a synapse at the end, which is related to another tangential process that also has a synapse at the end (see Fig. 2.6). In other words, many of these processes act functionally as both a dendrite and an axon, depending on whether a naturally activated electrical potential spreads toward or away from the cell body. This functional arrangement is possible because there are "reciprocal synapses" at the points where two processes from different motoneurons are connected.

Neural processes that normally conduct in either direction, and establish what amount to reciprocal synapses with like processes, were placed into a separate functional category, amacrine processes, by Cajal. He did this to distinguish them from dendrites, which normally transmit information toward the axon, which, in turn, normally transmits information away from dendrites and the cell body. According to this scheme, neurons can have three functionally distinct types of process: axonal, dendritic, and amacrine.

The simplest way to think about the functional significance of hydra nerve net architecture is to imagine that a stimulus applied to any one part of the animal will cause neural activity to spread in all directions through the net from the point of stimulation, and the strength of activity will decrease with distance from the point of stimulation because conduction tends to be decremental in amacrine processes. That is, in a nerve net stronger stimuli will spread farther, and produce greater responses, than weaker stimuli. It is easy to imagine that activity initiated by food on or near the tip of a tentacle could spread down the tentacle, producing the paddling motion responsible for bringing food toward the mouth—and that "better" food (in the sense of a stronger stimulus) might lead to more vigorous paddling.

Even in hydra, it is an exaggeration to say that neurons of the nerve net are uniformly distributed throughout the body. In fact, there is some increased density of cell bodies—a consolidation or differentiation of the nerve net—in regions such as the foot, mouth, and base of tentacles, where rudimentary "nerve rings" may be distinguished. These rings are specialized to control specific functions, such as the diameter of the mouth or tentacle movement during feeding and locomotion, and they are more obvious in more complex Cnidaria such as jellyfish.

While nerve nets with extensive amacrine processes seem to have appeared at the earliest stages of nervous system evolution, they have survived throughout the animal kingdom. For example, they are found in restricted, critical parts of the human brain such as the retina (amacrine cell layer) and olfactory bulb (granule cell layers) and in the human enteric nervous system (lining of the alimentary tract).

INTERNEURONS *Sign Switchers and Pattern Generators*

A third fundamental neuron type—in a sense, a third layer or stage—may be found in the vaguely differentiated nerve rings of hydra: the interneuron (Fig. 2.8). By definition, interneurons are neither sensory nor motor neurons; on connectional grounds they lie in between. There is a seemingly infinite variety of interneurons. However, they can be divided for the sake of convenience into two broad classes: local interneurons, with an axon that ramifies entirely within the region or cell group that generates it; and projection interneurons, with an axon that may ramify locally but always projects to another region or cell group.

Adding a third and final neuron type, with more and more varieties in more and more complex animals, provides animals with even more adaptive advantages, mainly in the realm of providing for increasingly sophisticated organization of neural circuitry—and thus behavior. A third level of control is added, and the potential for

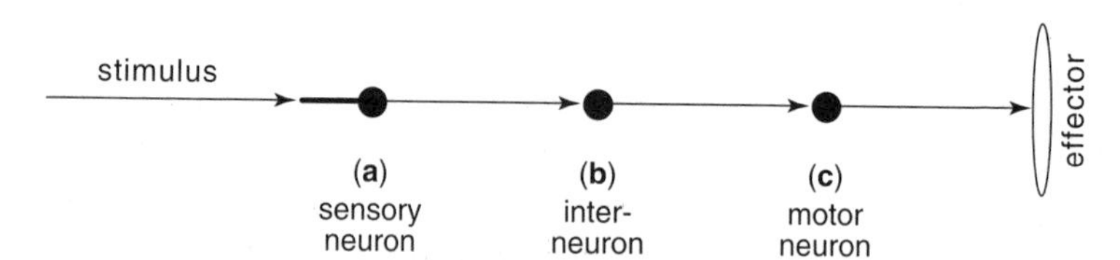

FIGURE 2.8 *Interneurons lie between sensory and motor neurons.*

divergence and convergence is expanded even further, probably exponentially.

But interneurons provide two additional features that are critically important for nervous system function, although they have not been discussed as yet: excitatory/inhibitory switching and pattern generation. In discussing chemical transmission between neurons (at synapses), we have assumed up till now that stimulation of a sensory neuron will lead to stimulation of a motor neuron—an excitatory input from sensory to motor neuron. However, transmission through neural circuits or networks involves both excitatory and inhibitory synapses. If one branch of a sensory neuron axon synapses directly on a motoneuron and another branch synapses on an inhibitory interneuron that innervates a second motoneuron, the first motoneuron will be excited whereas the second will be inhibited. Thus, in this simple example, an inhibitory interneuron acts as a switch from excitation to inhibition in the circuitry. In its simplest form, this circuit produces central excitation and lateral inhibition. In addition, the inhibitory interneuron just discussed may be thought of as a simple pattern generator. When the sensory neuron under consideration is excited, the first motoneuron is excited and produces a response whereas the second motoneuron is inhibited and thus does not produce a response. This is a stereotyped pattern of behavior (admittedly very simple) elicited by stimulating a sensory neuron.

The dynamics of nerve nets, and of neural circuits or networks in general, are even more interesting. Many sensory, motor, and interneurons appear to display spontaneous neural activity, so that, if left alone, they are almost always capable of producing a pattern of electrical impulses. Because of this, excitatory synapses tend to increase neuronal firing patterns, whereas inhibitory neurons tend to decrease these firing patterns. One illuminating consequence of this spontaneous "background" electrical activity is that excitation of motoneurons tends to increase muscle contraction, whereas inhibition of motoneurons tends to decrease muscle contraction (that is, it relaxes muscles). In short, synaptic inputs tend to modulate the firing rate of a neuron around some "spontaneous" or baseline rate, or set-point.

The spontaneous, intrinsic, activity of nervous systems is a profound concept because it invalidates the behaviorist view of animals that was fashionable in the first half of the twentieth century—the view that animals passively wait for external stimulation to trigger behavior in a purely reflex way. Quite to the contrary, the nervous system is spontaneously active, the nervous system is alive, and its activity is simply modulated—not controlled entirely—by external stimulation.

As mentioned, the structural organization of interneuron processes allows them to form pattern generators within neural circuitry. In addition, however, their spontaneous activity can be used in truly ingenious ways—for example, to generate spontaneous rhythmical activity patterns (as in the tentacles), or even as "biological clocks" in many animals.

OVERVIEW *Evolution of Architecture, Not Building Blocks*

Neurons first appeared during evolution in the Cnidaria, and their basic structure and function have stayed remarkably constant throughout the rest of the animal kingdom, including in the human brain. Neurons in all animals can be divided into three fundamental types: sensory, motor, and interneurons. Generally speaking, information is transmitted along neurons via electrical impulses associated with the plasma membrane, whereas it is usually transmitted between neurons and other cells via chemical synapses that use a mixture of neurotransmitters (although some *electrical synapses*—ephapses—are known). What has evolved dramatically is the complexity of nervous system organization, not its individual units or neurons.

READINGS FOR CHAPTER 2

Barrington, E.J.W. *Invertebrate Structure and Function*, second edition. Wiley: New York, 1979. This may be the best place to start if you want to understand the fundamental principles of biology.

Brusca, R.C., and Brusca, G.J. *Invertebrates*. Sinauer: Sunderland, 1990. Here is another excellent overview of invertebrate structure and function.

Bullock, T., and Horridge, G. *Structure and Function of the Nervous System of Invertebrates*, 2 vols. Freeman: San Francisco, 1969. This is a tour de force of scholarship and book production.

Gould, S.J. *Wonderful Life: The Burgess Shale and the Nature of History*. Norton: New York, 1989. He provides a fascinating account of how biological diversity may have evolved explosively, along with a love story about structural biology.

Grimmelikhuijzen, C.J.P., Leviev, I., and Carstensen, K. Peptides in the nervous systems of Cnidarians: structure, function, and biosynthesis. *Int. Rev. Cytol.* 167:37–88, 1996. An example of how neurotransmission is similar throughout the animal kingdom.

Hinkle, D.J., and Wood, D.C. Is tube-escape learning by protozoa associative learning? *Behav. Neurosci.* 1:94–99, 1994. No, but important to think about carefully.

Kandel, E.R., Schwartz, J.H., and Jessell, T.M. *Principles of Neural Science*, fourth edition. McGraw-Hill: New York, 1999. This is the standard introductory textbook.

Nakagaki, T., Yamada, H., and Tóth, A. Maze-solving by an amoeboid organism. *Nature* 407:470, 2000.

Oami, K. Distribution of chemoreceptors to quinine on the cell surface of *Paramecium caudatum*. *J. Comp. Physiol. A* 179:345–352, 1996. This is a nice introduction to what is being learned about the "sensory" capabilities of unicellular organisms.

Parker, G.H. *The Elementary Nervous System*. Lippincott: Philadelphia, 1919. A genuine classic and a delight to read.

Sakaguchi, M., Mizusina, A., and Kobayakawa, Y. Structure, development, and maintenance of the nerve net of the body column in hydra. *J. Comp. Neurol.* 373:41–54, 1996. The net is an amazingly dynamic system; are there molecular lessons for regeneration in our brains?

Swanson, L.W. Histochemical contributions to the understanding of neuronal phenotypes and information flow through neural circuits: the polytransmitter hypothesis. In: K. Fuxe, T. Hökfelt, L. Olson, D. Ottoson, A. Dahlström, and A. Björklund (eds.) *Molecular Mechanisms of Neuronal Communication*. Pergamon Press: New York, 1996 pp. 15–27. How are neuronal cell types best defined, and what is the logic behind neurotransmitter distribution patterns in neural circuits?

Szathmáry, E., and Smith, J.M. The major evolutionary transitions. *Nature* 374:227–232, 1995. What is the relationship between evolution and complexity?

Weiss, P. Autonomous versus reflexogenous activity of the central nervous system. *Proc. Amer. Phil. Soc.* 84:53–69, 1941. Brilliant.

Zigmond, M.J., Bloom, F.E., Landis, S.C., Roberts, J.L., and Squire, L.R. (eds.) *Fundamental Neuroscience.* Academic Press: San Diego, 1999. This is another excellent introductory textbook.

3

Centralization and Symmetry

Ganglia and Nerves

Without the relevant unifying concepts, comparative neurology becomes no more than a trivial description of apparently unrelated miscellaneous and bewildering configurational varieties, loosely held together by a string of hazy "functional" notions.

—HARTWIG KUHLENBECK (1967)

So far, we have considered protozoa and sponges, unicellular organisms and multicellular animals without a nervous system, along with the simplest animals with a nervous system—the jellyfish, corals, sea anemones, and hydra that have a more or less diffuse nerve net. All of these organisms either lack symmetry or are radially symmetrical, and their bodies are so simple that they lack clearly differentiated tissues. These features change dramatically when we come to what is commonly regarded as the next major branch of the evolutionary tree, the flatworms (phylum Platyhelminthes).

FLATWORMS *Bilaterally Symmetrical Predators*

These flat, unsegmented worms swim forward through the water very efficiently in search of food or a mate or to escape predators. Not surprisingly, the front end of the animal, which technically is called the rostral (for the Latin *rostrum,* or beak) end, contains specialized sense organs for detecting and identifying objects that the animal approaches as it swims through the environment. Now, for the first time, we encounter a bilaterally symmetrical body plan, with a longitudinal midline that divides it into right and left sides and extends from rostral to caudal (for *cauda* or tail) end. In addition, because the body is flat, there is a very clear top and bottom, technically referred to as dorsal (for *dorsum* or back, the "top" in this case) and ventral (for *venter* or belly, the "bottom" in this case) surfaces. Rostral/caudal and dorsal/ventral are the basic directional terms used to describe positional or topological relationships in all bilaterally symmetrical animals, including humans, although actual geometrical relationships often lead to confusion (see Appendix A). They are analogous to the north–south and east–west pointers on maps of the earth. In other words, there are perpendicular rostrocaudal and dorsoventral axes. A third axis, the mediolateral axis, is perpendicular to the first two and completes the scheme. *Medial* indicates a position toward the midline, and *lateral* indicates a position in the opposite direction.

A glance at Figure 3.1 reveals a strikingly organized nervous system in flatworms, compared to the nerve net in Cnidaria. It is immediately obvious that there is a massive condensation of neural elements into series of longitudinally (rostrocaudal) and transversely (mediolateral) oriented cords and nerves. There are also major condensations of nervous tissue in the rostral, "head," region of the animal, which are presumably associated with specialized sensory and motor mechanisms in the part of the animal that scans the oncoming environment during swimming.

The trend toward condensation of neural tissue is referred to as *centralization*, and it involves the aggregation of both axons (or nerve fibers), and neuronal cell bodies (also referred to as *somata*, or sometimes *perikarya*—actually the cell body minus its nucleus). In flatworms, a more or less pure condensation of nerve fibers is usually referred to as a *nerve*; a collection of neuronal cell bodies is called a *ganglion*, and a major, complex mixture of fibers and cell bodies is called a *nerve cord*. In the nineteenth century, the distinguished English philosopher Herbert Spencer and then Cajal used elaborate ar-

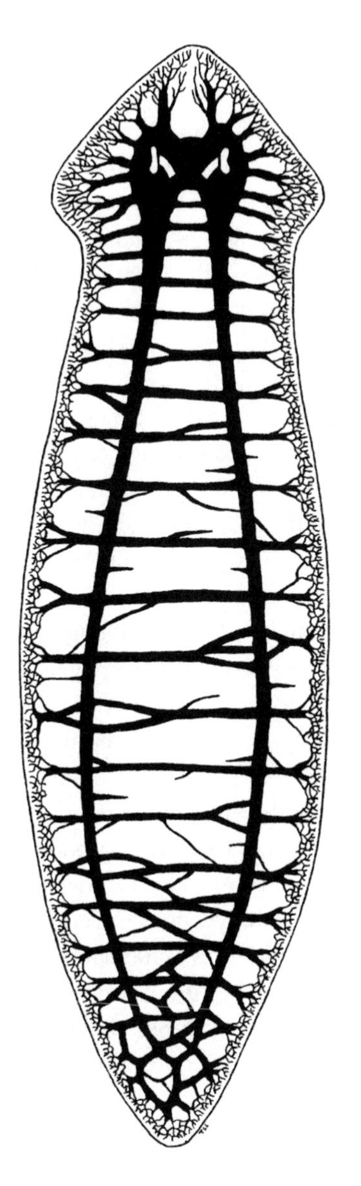

FIGURE 3.1 *The basic architecture of the flatworm (planarian) nervous system is shown from above in this drawing. There are two longitudinal nerve cords with a cerebral ganglion in the head region (at the top) and numerous transverse nerve cords. Reproduced with permission from T.L. Lentz,* Primitive Nervous Systems *(Yale University Press: New Haven, 1968, opposite p. 73).*

guments to show convincingly that centralization leads to *(a)* a conservation or more efficient use of biological material in the construction of neural circuits, *(b)* shorter distances traversed by neuronal processes to accomplish topologically similar circuit connections, and consequently *(c)* all things remaining equal, shorter conduction times for electrical impulses. These efficiencies of material, distance, and time would seem to be an almost inevitable consequence of evolution over immense periods of time and were foreshadowed by Dante's immortal aphorism, *Omne superfluum Deo et Naturae displiceat* (Everything superfluous is displeasing to God and Nature). Spencer had the distinction of anticipating Darwin in his famous argument that, as a general rule, the entire scope of organic evolution is accompanied by a change from the homogeneous to the heterogeneous and from the simple to the more complex.

More careful study of Figure 3.1 shows that the nervous system outlines the basic organization of the body. At the rostral end of the animal there is a head, and each side contains a huge mass of neurons that is pretty much bilaterally symmetrical and connected by an isthmus across the midline. This mass is the worm's brain, and its proliferation is associated with a concentration of special sensory and motor systems in the head. The successive differentiation of the rostral end of the body in evolution is referred to as *cephalization*.

An obvious, thick nerve cord extends down the right and left halves of the body from each lobe of the brain, and these two cords merge (or anastomose) near the tail of the animal. Because of their sheer volume, the brain and two longitudinal nerve cords in the body form the central part of the nervous system in flatworms. Nothing remotely as differentiated or condensed as this arrangement is found in the radially symmetrical, relatively sluggish Cnidaria with more or less condensed nerve nets.

Sensory information about the environment in front of the head is gathered by various receptors and then processed in the brain, which, in turn, controls swimming behavior by sending commands down the nerve cords. This swimming behavior is triggered by waves of muscle contraction passing along the animal's body, and the precise timing of these waves determines the route taken by the animal and the speed used along it. The sequential activation of muscles along the body is actually coordinated by information that is distributed by a ladder-like arrangement of thinner, transversely arranged nerves, which also carry sensory information from the body to the thicker longitudinal cords. Where the transverse nerves cross the midline they are called *commissures*, which obviously allow information to pass from one side of the body to the other. This is the peripheral part of the worm's nervous system, as opposed to the central part.

It is immediately obvious that the flatworm's brain, cords, and nerves form an anastomotic network or reticulum when viewed with the naked eye. However, just as in hydra, the actual circuitry of this nervous system is formed by tiny individual neurons that are linked together by chemical synapses or switches in very complex networks—it is not a cellular reticulum. The actual course of axons from specific neurons or groups of neurons within and through the ladder-like arrangement of anastomotic cords, nerves, and brain is an exceptionally difficult problem to solve, with plenty still to be learned.

As neural elements condense and centralize during the course of evolution, they tend to form two broad classes of structure: collections of neuronal cell bodies that are called *ganglia*, and bundles of axons referred to as *nerves*. In the case of the central part of the nervous system it is becoming unfashionable to refer to ganglia. Instead, it is preferable to reserve the term ganglion for a cluster of nerve cell bodies in the peripheral nervous system. So nowadays, for example, the "cerebral ganglion" is thought of as the brain, and at least in a very general way it has the same behavioral control function in all animals, from planaria to people.

It is within the ganglia and central nervous system that most synapses between neurons occur. In the ganglia of invertebrates, nerve cell bodies tend to lie around the outside or periphery, and most synapses, which are axodendritic, occur in the center, where the tissue is called *neuropil* (a complex mixture of axons, dendrites, and synapses—and few if any cell bodies). Central parts of the invertebrate nervous system also have considerable regions of neuropil, although its arrangement is much more irregular and complex than in typical ganglia.

In flatworms, and in all other invertebrates except the Cnidaria, most interneurons and motoneurons are *unipolar* (Fig. 3.2). They have an ovoid cell body, a single massive process that extends for a longer (in, for example, motoneurons and projection interneurons) or shorter (in, for example, local circuit interneurons) distance, and a multitude of thin dendrites that extend transversely from the proximal region (closest to its origin from the cell body) of the axon. This is in stark contrast to vertebrates where, instead, the vast majority of interneurons and motoneurons are multipolar, with one or more dendrites extending from the cell body and an axon that arises from the cell body or one of the dendrites.

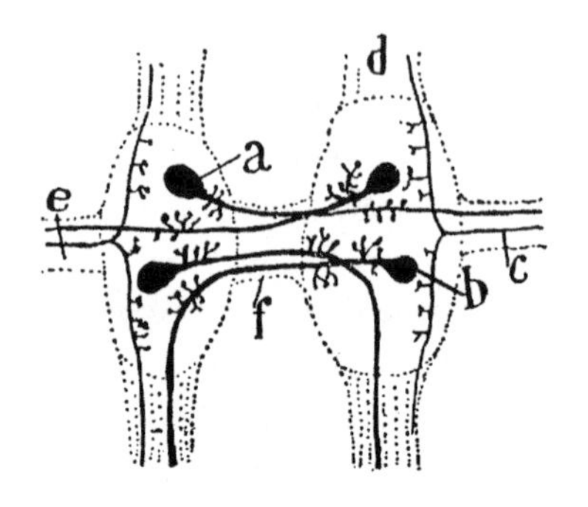

FIGURE 3.2 *The typical appearance of unipolar neurons in invertebrate ganglia is shown here.* Key: *a, crossed motor neuron; b, commissural or sensory association neuron; c, sensory fiber from the integument; d, longitudinal commissure; e, nerve containing centrifugal motor fibers and centripetal sensory fibers; f, transverse commissure. From S.R. Cajal,* Histologie du système nerveux de l'homme et des vertébrés, *vol. 1 (Maloine: Paris, 1909).*

The fact that most invertebrate synapses are formed in an incredibly fine-grained neuropil, whereas most vertebrate synapses are formed on the massive dendrites of neurons that are easy to identify, led Cajal to conclude that neural circuitry is much easier to elucidate in vertebrates than in invertebrates. This was based on his valiant though personally disappointing attempt at the neuroanatomy of the ant after publishing his masterpiece on the histology of the entire vertebrate nervous system.

SEGMENTED WORMS *Internal Ventral Nerve Cord*

Earthworms and leeches are typical examples of the some 15,000 species of segmented worms in the Annelida phylum, which has a more differentiated body plan than the simple flatworms. In annelids the nervous system has become even more centralized. As in flatworms, there is a dorsal brain (sometimes called the *suprapharyngeal ganglia,* because they lie dorsal to the innermost end of the mouth segment of the digestive tract). However, in the segmented worms, and in all other invertebrates, the major longitudinal nerve cords come to lie next to each other ventral to the gut, or they fuse into a single ventral nerve cord in that location (Fig. 3.3).

The composition of the ventral nerve cord in segmented worms is interesting, and its embryological origin seems to differ fundamentally from the analogous spinal cord in vertebrates. As the name implies, one basic way segmented worms differ from flatworms is that

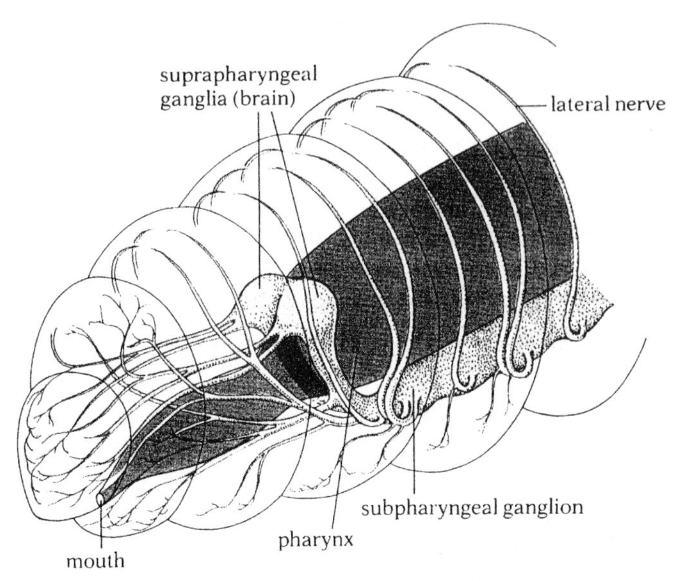

FIGURE 3.3 *The basic arrangement of the nervous system in the rostral end of the earthworm (annelid) is shown in this drawing. Relative to the alimentary canal (black), the brain is dorsal and the nerve cord formed by the subpharyngeal ganglia is ventral. Note that the distribution of peripheral nerves is more complex in the rostral tip ("head region") of the animal. Reproduced with permission from J.L. Gould and W.T. Keaton,* Biological Science, *sixth edition (Norton: New York, 1996, p. 1001).*

much of their body length is formed by the serial repetition of a transverse unit called a *segment.* The basic idea is that this is a genetically efficient way to program the development of a more complex animal because essentially the same genetic program can be used over and over—in each segment or metamere.

Each body segment in an earthworm, for example, has a ventral ganglion and paired (right and left) nerves that circle dorsally and ventrally to innervate various parts of the segment. In the adult, these segmental ganglia more or less fuse and are bound into a cord by the presence of innumerable longitudinally oriented nerve fibers (axons). The organization of neuronal interconnections in this ventral nerve cord is even more complex than it is in the ganglia discussed for the flatworms—implying that more complex behaviors are mediated by this circuitry (compare Figs. 3.2 and 3.4).

MORE EVOLVED INVERTEBRATES

There are many phyla of invertebrates that have a more highly differentiated body plan than the flatworms and segmented worms: insects, crustaceans, mollusks, echinoderms, and so on. At one end of the spectrum is the tiny fruit fly, which has been the favorite of neurogeneticists for a century, and at the other is the giant octopus, with a brain that at least superficially puts many vertebrates to shame in terms of sheer size and complexity (Fig. 3.5). Nevertheless, they all

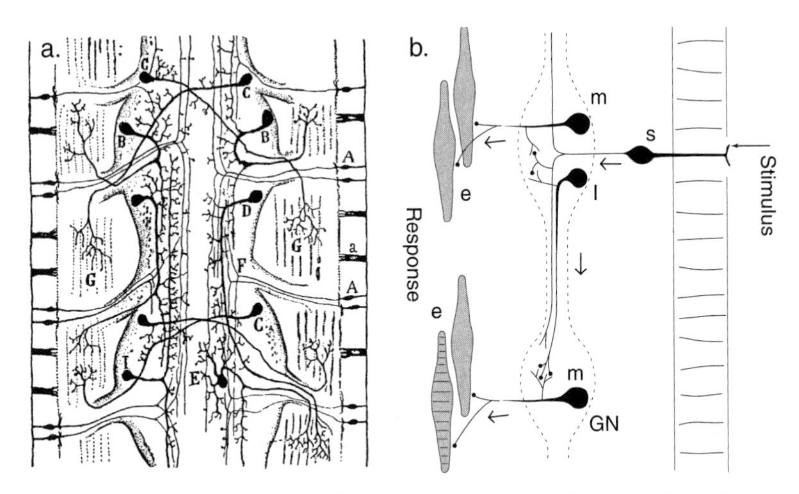

FIGURE 3.4 *Some features of earthworm nervous system organization are shown here as viewed from above. (a) This drawing is based on Golgi-impregnated tissue and shows both the right and left sides of the body. (b) This more schematic drawing shows only the right side of the animal. Key: Arrows indicate direction of information flow; a, epithelial cell; A, sensory cells of the skin; B, ipsilateral motor cells within the central ganglia; C, motor cells with crossed processes; D, motor cells with longitudinal ipsilateral processes; e, effectors; E, multipolar motor cells; F, sensory axon bifurcation; G, terminal ramifications of a motor neuron on a muscle; GN, ganglion; I, interganglionic association cells; m, motor neuron; s, sensory neuron. Part a is from S.R. Cajal,* Histologie due système nerveux de l'homme et des vertébrés, *vol. 1 (Maloine: Paris, 1909).*

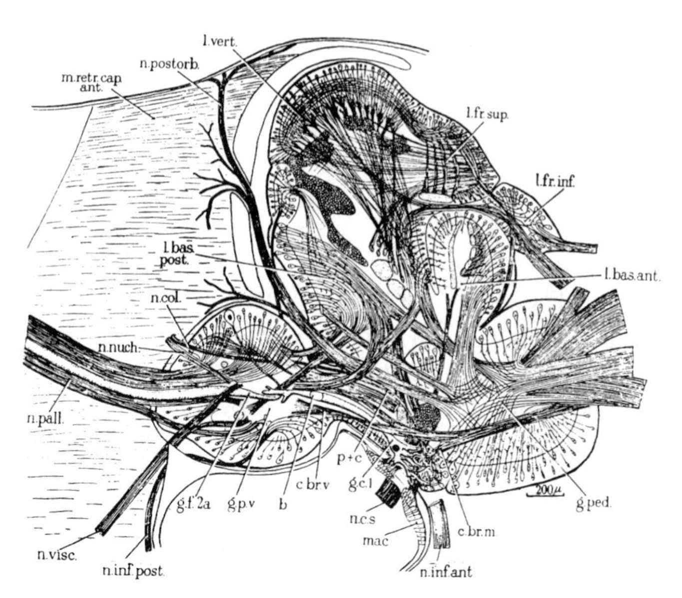

FIGURE 3.5 *The brain of a young octopus as viewed in a parasagittal section, with rostral to the right. The octopus brain contains on the order of 150 million neurons. Reproduced with permission of the Royal Society from J.Z. Young, Fused neurons and synaptic contacts in the giant nerve fibres of Cephalopods, Phil. Trans. R. Soc. B, 1939, vol. 229, p. 471.*

have a dorsal brain and a ventral nerve cord that lies between the gut and the ventral body wall. It is remarkable that the brain and ventral nerve cord of invertebrates are derived from the ectodermal layer of the embryo—the layer where the sensory and motor neurons of the hydra nerve net are also generated. During embryogenesis, these ectodermally derived neurons migrate into the interior of the animal (that is, into the mesodermal layer), in a process that is called *delamination.*

OVERVIEW *Polarity, Regionalization, Bilateral Symmetry, Segments*

Before delving into the vertebrate nervous system, let us pause for a moment to consider what we have learned about the biology of the "simpler" organisms. First, recall that even unicellular organisms such as protozoa display three classes of remarkably sophisticated behaviors that are, in fact, common to all "higher animals" because they are essential for survival: ingestive (or appetitive), defensive, and reproductive.

Second, the most primitive multicellular animals (sponges) have no nervous system and yet display the same three classes of behavior. These animals evolved different classes of cells that are specialized for particular tasks; they have seized on the advantages associated with the division of labor principle. The myocyte is one of these cell types, and it is able to contract when it is directly stimulated. The activity of these independent effectors allows the animal to regulate various behaviors much more effectively than do the protozoa.

Third, the radially symmetrical Cnidaria display a new cell type, the neuron, which is arranged in a more or less diffuse nerve net that now controls the activity of the (former) independent effectors in a much more effective way, allowing even more sophisticated behaviors. The fundamental morphology, physiology, and chemistry of individual neurons, as well as their mode of functional contact with other cells, is remarkably similar in all animals.

And fourth, the evolution of bilateral symmetry in animals (initially in worms) is associated with *(a)* localized condensations or centralization of the nervous system into ganglia, nerves, and nerve cords; *(b)* polarity in the sense that there is a head at one end and a tail at the other end; *(c)* regionalization of the nervous system such that there is a highly differentiated brain in the head, as well as nerve cords in the body and tail; and *(d)* segmentation of the nervous system and rest of the body. As with nerve nets, nerves, ganglia, and nerve cords are found in the "higher" animals. In humans, nerves and ganglia are the principle components of the peripheral nervous system, and the great sympathetic chains are bilateral nerve cords.

READINGS FOR CHAPTER 3

Barrington, E.J.W. *Invertebrate Structure and Function*, second edition. Wiley: New York, 1979.

Breidbach, O., and Kutsch, W. (eds.) *The Nervous Systems of Invertebrates: An Evolutionary and Comparative Approach.* Birkhäuser: Basel, 1995. It has reviews by a panel of experts of recent detailed research on selected groups of invertebrates.

Brusca, R.C., and Brusca, G.J. *Invertebrates*. Sinauer: Sunderland, 1990.

Bullock, T., and Horridge, G. *Structure and Function of the Nervous System of Invertebrates*, 2 vols. Freeman: San Francisco, 1969.

Cajal, Santiago Ramón y. *Histologie du système nerveux de l'homme et des vertébrés*, 2 vols. Translated by L. Azoulay. Maloine: Paris, 1909, 1911. For American translation, see N. Swanson and L.W. Swanson, *Histology of the Nervous System of Man and Vertebrates*, 2 vols. Oxford University Press: New York, 1995. Chapter 1 has a brilliant essay on the basic plan of the nervous system; the rest is a goldmine of neuroanatomical knowledge (much of it still valid) from around the turn of the nineteenth century.

Kuhlenbeck, H. *The Central Nervous System of Vertebrates:* Vol. 2, *Invertebrates and the Origin of Vertebrates*. Karger: Basel, 1967. It is a stimulating contrast to Bullock and Horridge.

Lentz, T.L. *Primitive Nervous Systems*. Yale University Press: New Haven, 1968. This is a nice overview.

Spencer, H. *Illustrations of Universal Progress*. Appleton: New York, 1890.

Strausfeld, N.J. *Atlas of an Insect Brain*. Springer-Verlag: Berlin, 1976. This is an eye-opening introduction to the structural organization of the fly brain.

Young, J.Z. *A Model of the Brain*. Oxford University Press: London, 1964. This is a brilliant attempt to understand the functional organization of the octopus brain—and brains in general.

4

The Basic Vertebrate Plan

Nervous System Topology

A diagram is a changing structure. It must be improved, now here, now there. Certain parts often need to be torn down and rebuilt. It has been contended that we ought not to make use of diagrams in a subject so full of gaps as is our knowledge of the structure of the nervous system. Let us rather hold, with old Burdach, who wrote in 1819, "The gathering together of material for the building is not all that is necessary. Every time that a new supply is obtained, we should renew our attempts to fit it into the building. By thus giving it a form the spirit of investigation is not hampered in its advance; on the contrary, it is when we first obtain a view of the whole that we see the gaps in our knowledge and learn the direction which our investigations must take in the future. May the attempts at this structure ever be renewed. No one who works at it but adds something to our knowledge."

—LUDWIG EDINGER (1891)

Every now and then science generates an idea that is widely viewed as profane or seditious—a view of the relationship between humans and the rest of the universe that subverts time-honored cultural traditions and yet in the long run is supported by the facts. The first great intellectual revolution along these lines, which has finally been won, was started in 1543 by the Polish astronomer and physician, Nicolaus Copernicus. In his book, *De Revolutionibus*, which was published the year that he died, a fundamental conclusion was that humans do not occupy a place at the center of the universe, as Aristotle and the Bible stated or at least seemed to indicate, respectively. Instead, we merely reside on a planet that circles around the sun. The latest revolution, which began in 1859 and is still being debated in the public's mind, was sparked by Charles Darwin's book on *The Origin of Species*. As we know, the immortal English naturalist had the courage to argue with dignity and force that humans are not the products of special creation but, instead, are the products of "chance" evolution from "lower" animals over an unimaginably long period of time.

In Chapter 1 it was pointed out that the seeds of evolutionary thinking can be traced back to antiquity. For example, Aristotle recognized that the great diversity of life forms can be accounted for by a small number of fundamental body plans, each with many variations on a common theme. However, this did not extend to thinking about you and me, and one can only imagine the effect that the drawing shown in Figure 4.1 had on the reflective public when it was published by Pierre Belon in 1555. He is considered by many to be the founder of modern comparative anatomy—the first great practitioner of that science since Aristotle—and he had the brilliant insight to render the bird skeleton at the same scale as the human skeleton, driving home his basic discovery that the skeleton of the two is essentially homologous. What does this mean?

Simple observation reveals that the bones of the head, neck, trunk, upper limbs (wings in birds, arms in humans), and lower limbs are strikingly similar if one ignores details of length, thickness, and exact shape. In other words, the topology of the parts—their relationship to each other—is basically the same in birds and humans, although the precise geometry of individual parts, and the overall geometrical appearance of the assembly, may be different. If one accepts this principle, it doesn't take much imagination to postulate that the muscular system of birds and humans (a mammal) is also homologous in basic organization, simply because muscles are attached to and move the skeleton. In fact, it might be reasonable to

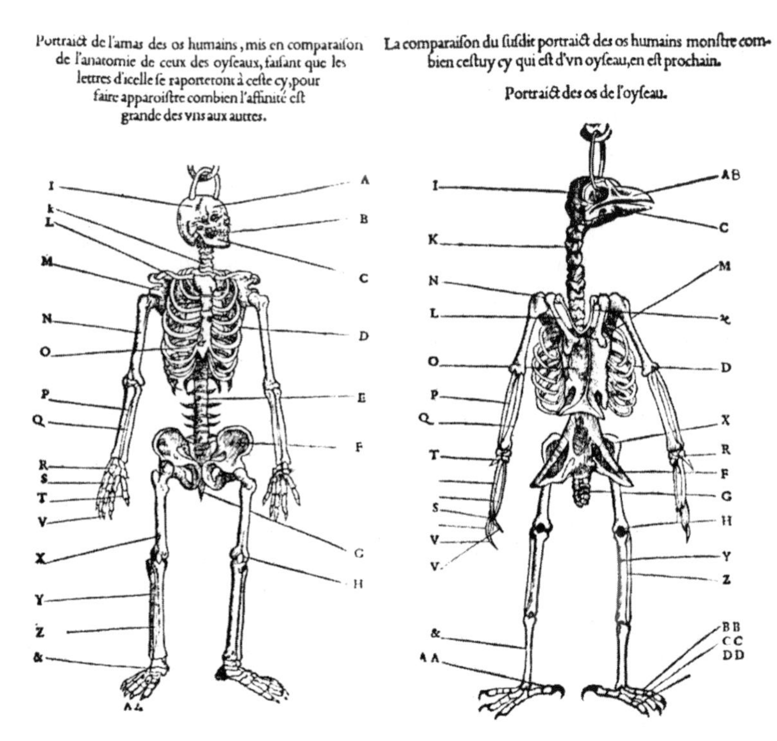

FIGURE 4.1 *This remarkable comparison of the human and bird (chicken) skeleton, where homologous bones are labeled identically, was published in 1555 by Pierre Belon, in his famous book,* L'histoire de la nature des oyseaux, avec leurs descriptions, & naïfs portraicts retirez du naturel.

postulate that all of the great systems of the body—including the nervous system—may share a fundamental architecture.

Results emerging from centuries of biological research after Belon have led to the broad generalization that the entire Vertebrata subphylum of Chordata (which includes fish, amphibians, and reptiles, in addition to birds and mammals) has the fundamental body plan shown in Figure 4.2. At the core, this body plan includes four elements that are displayed at some time during the life cycle of all of them. First, there is a notochord, which is a stiff rod that extends

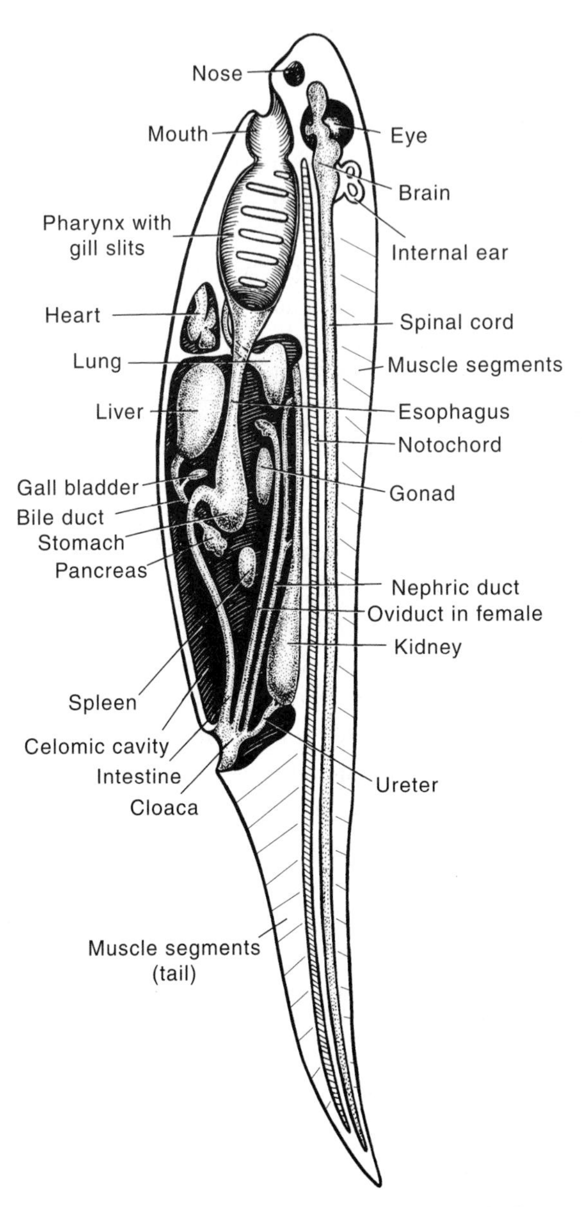

FIGURE 4.2 *The basic vertebrate body plan is shown in this schematic diagram. Reproduced with permission of Brooks/Cole, an imprint of the Wadsworth Group, a division of Thompson Learning, from A.S. Romer,* The Vertebrate Body, *fifth edition (Saunders: Philadelphia, 1977, p. 3).*

along the midline of the body, preventing it from shortening or bending excessively during locomotion (swimming or walking), which is a very efficient arrangement. Second, there is a dorsal central nervous system, which is hollow and consists of a large brain in the head and a thinner spinal cord extending down the trunk. Recall that invertebrates have a condensed ventral nerve cord and no notochord (Chapter 3). Third, vertebrates have a series of pharyngeal or branchial arches (paired gill slits in fish and amphibians) for respiration and feeding that are associated with a ventral, unidirectional gut for digestion (which starts in the mouth and ends in the anus, or cloaca). And fourth, vertebrates have segmented skeletal muscles along the body and in a postanal tail. Early in evolution virtually all of the skeletal muscle was used for swimming, but in land animals the limbs and head became much more differentiated and mobile.

Although fewer than 0.1% of the animal species on earth today incorporate the vertebrate body plan, those that do show by far the most complex, modifiable behavior in the animal kingdom, and they are certainly of greatest interest to us. In the end, humans are really just a specialized vertebrate, and the organization of our nervous system—with its all-important brain—is simply a reflection of how the rest of our body is specialized relative to other vertebrate classes and species. It is a variation on the basic vertebrate plan, and we shall now examine that plan as revealed through the embryological development of the nervous system. The embryo starts out simple (a single cell, the fertilized egg) and becomes more differentiated over time, just as the nervous system underwent progressive differentiation during the course of evolution (Chapters 2 and 3). This is a venerable approach in biology—the comparison of ontogeny (individual development) and phylogeny (species evolutionary history)—because each approach proceeds from simple to complex.

EMBRYOLOGICAL PERSPECTIVES

As we saw in Chapter 1, Aristotle realized that the embryo would reveal the fundamental body plan as one follows over time its progressive differentiation and specialization. It goes without saying that he did not have at his disposal the tools needed (like a microscope) to carry this line of research very far. Aristotle's first great successor in this arena was the Italian anatomist Marcello Malpighi, who was born in 1628 and died in 1694. He was a Fellow of the Royal Society of London and physician to the pope, and his embryological masterpieces on the development of the chick embryo were published in the mature phase of his career, in 1673. For this research he used simple compound microscopes, and he discovered that the earliest recognizable shape of the nervous system is a regionalized, spoon-shaped plate, which is followed in the brain region by a stage with three longitudinally arranged swellings and then a stage with five longitudinally arranged swellings (Fig. 4.3). At the earlier plate stage, the broad region is near the rostral end of the embryo and corresponds to the future brain, whereas the narrower stem lies more caudally and corresponds to the future spinal cord. Malpighi noted that although the brain region goes on to form a series of swellings, the spinal cord region retains a simpler, smoother, narrower shape.

Nothing of any real significance was added to Malpighi's account for a century and a half—until the landmark work of Karl von Baer was published between 1828 and 1837. Baer provided the first truly full and adequate description of chick development, but much more importantly, he followed this with a masterly elaboration of laws that govern vertebrate development in general. In doing so, he laid out essentially everything we know today about the macroscopic features of development—that is, features not based on the cell theory, which was not articulated until 1839, by Matthias Schleiden and Theodor Schwann.

The story that Baer told is fundamental. At an early stage of differentiation, the vertebrate embryo is trilaminar—it consists very simply of three stacked sheets that are roughly oval in shape. They are ectoderm dorsally, endoderm ventrally, and mesoderm in the middle. Furthermore, the neural plate is a bilateral, midline differentiation of the ectoderm, the layer that goes on to form the outer surface (skin) of the animal. During later stages of development, the

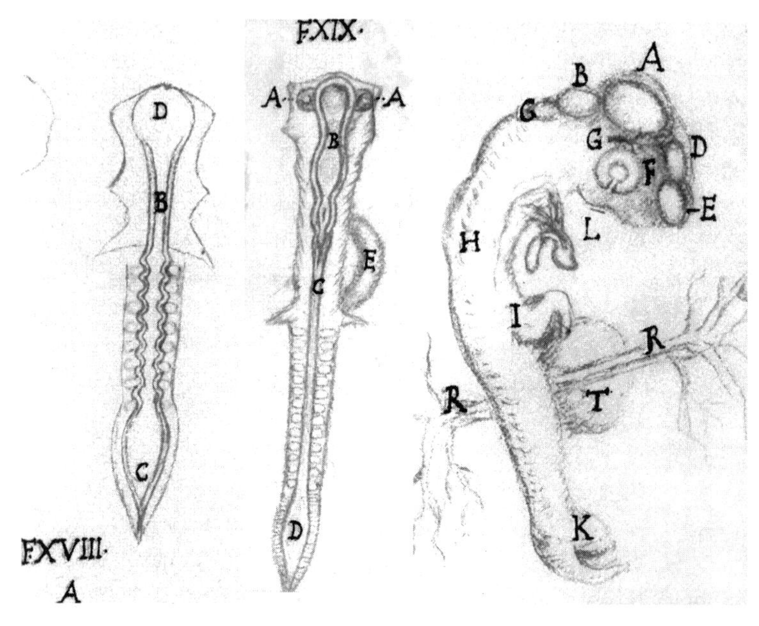

FIGURE 4.3 *These drawings by Marcello Malpighi were published in 1673 and illustrate early development of the chick central nervous system. In the drawing on the left, he shows the neural plate with prospective brain (D) and spinal cord (B) regions, and below that an indication of eight somites. The middle drawing shows the three-vesicle stage of the neural tube (B) and the eyes (A). The drawing on the right shows the five-vesicle stage, with what are now called paired endbrain vesicles (E), interbrain vesicle (D), optic cup and choroid fissure (F), midbrain vesicle (A), and hindbrain vesicle differentiated into pons (or metencephalon, B) and medulla (or myelencephalon, C).*

three embryonic ("germ") layers roll into a tube and fuse ventrally, so that the surface ("somatic") ectoderm is on the outside and the endoderm is on the inside, lining the gut. The mesoderm goes on later to form bone, muscle, blood vessels, and other tissues.

To state this more formally, Baer identified three broad stages of differentiation in all vertebrate embryos. First, there is primary differentiation or formation of three embryonic layers, the trilaminar disc stage, that ends when they, in turn, form concentric tubes. Next, there is secondary, histological differentiation within the layers. Finally, there is tertiary, morphological differentiation of primitive organs, most of which he described quite adequately. In short, Baer showed how the basic plan of the vertebrate body illustrated in Figure 4.2 is constructed from three stacked sheets in the early embryo. His fundamental conclusion—that in embryogenesis general features appear before specialized features—is perfectly captured in an anecdote from the first volume of his masterpiece (1828): "I have two small embryos preserved in alcohol that I forgot to label. I cannot at the moment determine the genus to which they belong. They may be lizards, small birds, or even mammals." Figure 4.4 makes the point

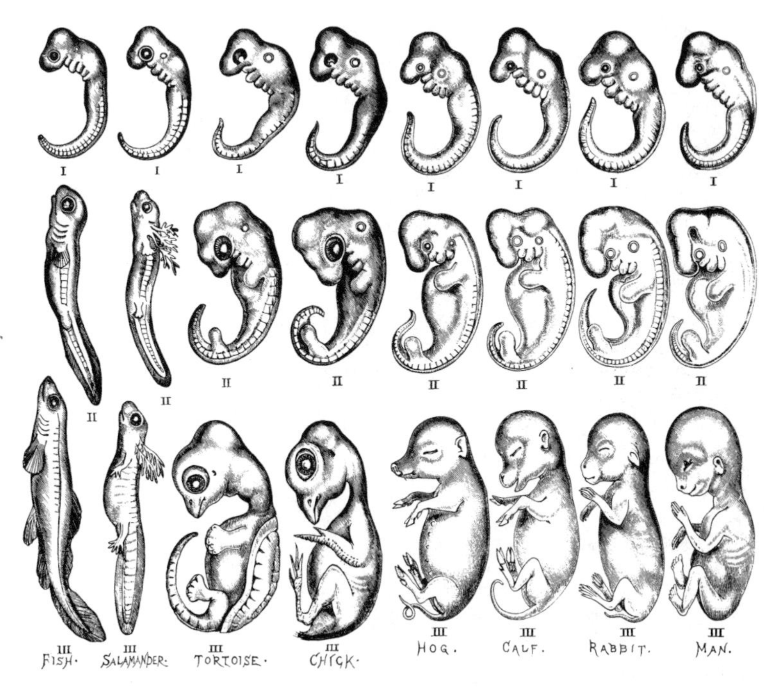

FIGURE 4.4 *Baer's law, that in vertebrate embryogenesis the general develops before the specific, is nicely illustrated here. Approximately 4-week* (top row: *I, three brain vesicle*), *5-week* (middle row: *II, five brain vesicle*), *and 8-week* (bottom row: *III*) *human embryos are shown. From G.J. Romanes,* Darwin, and After Darwin *(Open Court: Chicago, 1901).*

graphically and brings new meaning to the adult vertebrate body plan first illustrated by Belon (see Fig. 4.1).

EARLIEST STAGES OF MAMMALIAN DEVELOPMENT

Let us now come back to Baer's stage of primary embryonic differentiation and sketch the fascinating story of how the trilaminar embryonic disc is actually formed in mammals, using the human as a typical example (Fig. 4.5). The power of this approach lies in its simplicity. The starting point is a single cell, the fertilized egg, which has one copy of DNA (the genetic program for development and mature

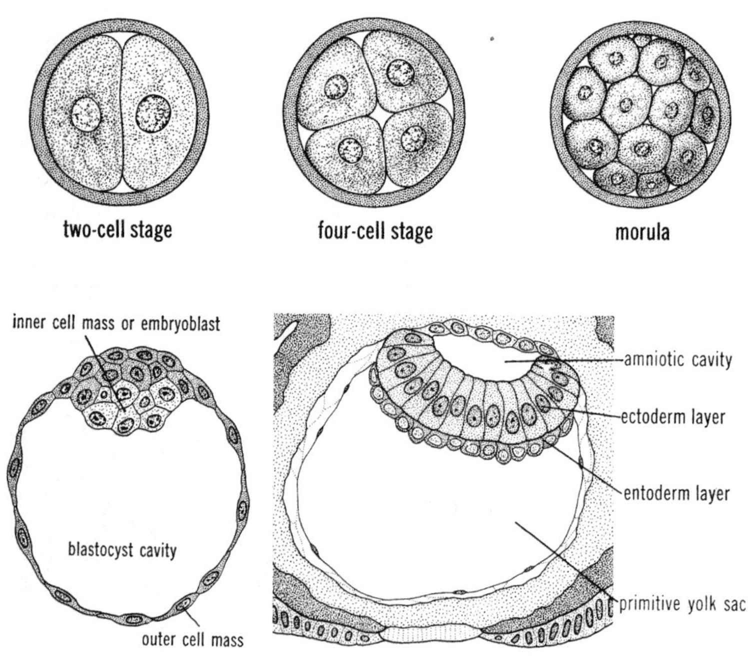

FIGURE 4.5 *Formation of the human bilaminar embryonic disc* (lower right) *from the two-cell stage* (upper left) *is shown in this series of drawings. The inner cell mass develops during the fourth day after fertilization, and the blastocyst* (lower right) *is about 9 days old. Adapted with permission from J. Langman,* Medical Embryology, *third edition (Williams and Wilkins: Baltimore, 1975, pp. 27, 29, 41).*

function) from the sperm and a different copy from the egg itself. The fertilized egg divides a number of times to form a ball of more or less similar cells known as a *morula*, and then a dramatic event occurs: a large, fluid-filled cavity develops within the morula. This cavity is known as the blastocyst cavity, and it goes on to form the yolk sac at later stages. In contrast, the cells themselves are arranged into a thin outer cell mass that lines the yolk sac, and a condensed inner cell mass becomes the prospective embryo itself.

A second cavity—the amniotic cavity—then develops within the inner cell mass. A remarkable thing happens at this (blastocyst) stage. The part of the inner cell mass between the two cavities becomes organized into two monolayers of cells: the bilaminar embryonic disc. The layer "on top," next to the amniotic cavity, is the ectoderm, and it is referred to by convention as *dorsal* in the embryo. Conversely, the layer "on the bottom," next to the primary yolk sac, is the endoderm, and it is referred to as *ventral* in the embryo at this stage of development.

Then comes the single most important event for nervous system formation. To set the stage, one must look down on the bilaminar disc from above and identify two clear features that divide the ectoderm into right and left halves. Near one end of the disc there is a circular patch where the ectoderm and endoderm appear to be fused or "welded" together, and at the other end there is a groove with a swelling on the end (Fig. 4.6). The circular patch, which lies near the rostral end of the embryonic disc, is rather boring—it is called the *oropharyngeal membrane*, and it will disappear to become the opening between the mouth and the pharynx (throat), the rostral end of the gut. In contrast, the groove and swelling lie caudally. They are called the *primitive streak* and *primitive node (of Hensen)*, and they are specialized regions of the midline ectoderm that do something remarkable: they generate cells that migrate between the ectoderm and endoderm to form the mesoderm.

The single most important event for nervous system formation is the differentiation of Hensen's node, which Hans Spemann and Hilde Mangold showed in the 1920s is a primary "organizer" of the nervous system. When the organizer is removed, the nervous system

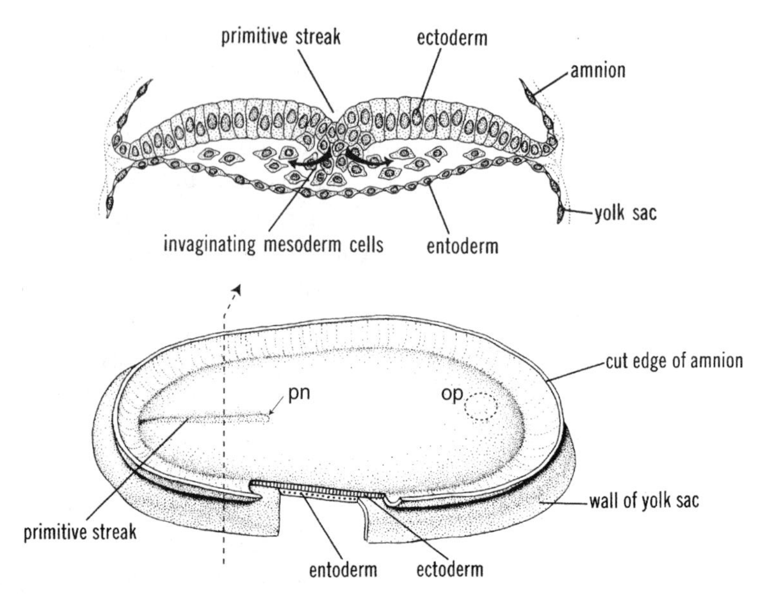

FIGURE 4.6 *The formation of the human trilaminar disc is illustrated here. The top figure is a cross-section through the dorsal view of the disc in the bottom half of the figure (where rostral is to the right). The dashed line through the primitive streak in the bottom figure shows the level of the cross-section. This stage of development is seen in the 16-day old embryo.* Key: *op, oropharyngeal membrane; pn, primitive node of Hensen. Adapted with permission from J. Langman,* Medical Embryology, *third edition (Williams and Wilkins: Baltimore, 1975, p. 51).*

fails to develop at all, whereas a transplanted organizer directs the construction of a second nervous system in an embryo. A molecular explanation of the organizer program is one of the holy grails of neuroscience that remains to be captured, although a rather minute description of cellular events associated with nervous system differentiation has been available for some time.

To begin with, the cells in the rostral tip of Hensen's node migrate rostrally along the midline, toward the oropharyngeal membrane, which provides a barrier to their further movement. These midline cells, which stretch from Hensen's node to the oropharyngeal membrane, become the mesoderm that forms the notochordal process. This process (despite the name, it is a morphological structure) is very interesting because it, in turn, forms the notochord, one of the defining characteristics of vertebrates. But the situation is even more interesting because some factor or combination of factors secreted from notochordal cells then diffuses dorsally to induce changes in midline regions of the overlying ectoderm—and these changes actually represent the induction of a dorsal central nervous system, another cardinal feature of vertebrates. The most thoroughly examined candidate factor to date is the protein *sonic hedgehog*, which is the vertebrate homolog of a protein discovered in the fruitfly *Drosophila* that is involved in segmentation of the body during embryogenesis. As the embryonic disc (which is now trilaminar because of the mesodermal layer) grows, cells are added to the caudal end of the notochordal process—that is, to the end adjacent to Hensen's node. In other words, the notochordal process grows from rostral to caudal, and the overlying neuroectoderm is induced from rostral to caudal. There is a temporal gradient in neuroectoderm formation, with the rostral end being the oldest and the caudal end the youngest.

NEURAL PLATE *Brain and Spinal Cord*

Now we can return to the neural plate stage discovered by Malpighi (Fig. 4.3, left) and describe its fate in cellular terms. To begin with, the inductive influence of the notochordal process is responsible, at least in part, for dividing the ectoderm into a midline, spoon-shaped neuroectoderm, and a peripheral somatic ectoderm, which goes on to form the skin. The two regions are easily distinguishable because the neuroectoderm is thicker than the somatic ectoderm so that, when examined in cross-section, cells in the neuroectoderm are taller than those in the somatic ectoderm and this is why the neural plate is easy to see (Fig. 4.7, right). As mentioned several times already, the bowl of the spoon (which is rostral and formed earliest) is the presumptive brain, whereas the handle (which is caudal and formed later) is the presumptive spinal cord, although there is no clear borderline between the two at this stage. In addition, there is a groove, the neu-

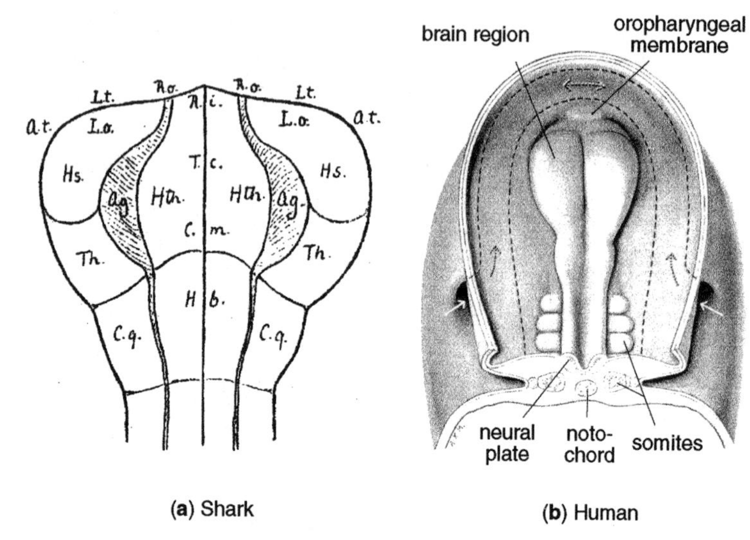

FIGURE 4.7 *Dorsal views of the neural plate (neuroectoderm) in shark (a) and human (b) embryos. The notation in (a) indicates Wilhelm His's fate map of the shark neural plate.* Key: *Ag., optic stalk; A.t., terminal angle; C.m., mammillary body; C.q., quadrigeminal body; Hb., "tegmental" eminence; Hs., pallial hemisphere; Hth., hypothalamus; L.o., olfactory lobe; L.t., terminal lamina; R.i., infundibular recess; R.o., optic recess; T.c., tuber cinereum; Th., thalamus; with the optic chiasm falling between the infundibular and optic recesses. Example (a) from W. His,* Arch. Anat. Physiol. Leipzig, Anat. Abth. *pp. 157–171, 1893. Example (b) adapted with permission from W.J. Hamilton, and H.W. Mossman,* Human Embryology, *fourth edition (Williams and Wilkins: Baltimore, 1972, p. 77).*

ral groove, that extends down the midline and separates the plate into right and left halves called *neural folds.*

In short, we have a polarized (rostrocaudal), bilaterally symmetrical, regionalized sheet of cells that represents the future central nervous system, with its brain and spinal cord regions. The cells form a monolayer of progenitor or stem cells that divide over and over to produce more progenitor cells at an exponential rate. The production of neurons occurs later, as we shall soon see.

Put another way, the architecture of the central nervous system at its earliest stages of development is incredibly simple. Topologically, it is a flat, bilaterally symmetrical sheet that is one cell thick. Naturally, there has been great interest in determining whether any of the later regional subdivisions of the nervous system can be detected in the neural plate. The first and probably most insightful attack on this problem was conducted by Wilhelm His, the greatest neuroembryologist of the nineteenth century. In examining his vast collection of vertebrate embryos, he believed that the shark neural plate displays enough features to distinguish the major subdivisions observed in the brain of other, more advanced, vertebrates at later stages of development. In this model, the rostral end of the neural plate is at the level of the infundibulum, the stalk of the pituitary gland that develops in the midline, just at the level of the oropharyngeal membrane (Fig. 4.7a). Without going into details now, simply note that His identified a series of longitudinal and transverse subdivisions in the shark neural plate.

In mammals, the first two differentiations of the neural plate that can be identified with certainty are in the brain division and include *(a)* a rostral region that goes on to form the retinas and optic nerves and *(b)* a caudal region that is associated with the inner ear. There is some evidence to suggest that, somewhat later, a midbrain region of the neural plate can be seen, which would imply that the region of the plate rostral to it would be forebrain and the region caudal to it would be hindbrain.

Currently, there is intense interest in determining whether there are regionalized patterns of gene expression in the mammalian neural plate. Such patterns might suggest molecular mechanisms for regionalization of the neural plate, and thus regional differentiation of neuroepithelial cells, before the actual production of neurons themselves and even before the neural tube is formed. Such patterns are beginning to surface, although their significance is still fairly obscure, partly because their relationships to the morphological features just mentioned are not entirely clear. We can expect great progress here in the very near future.

NEURAL TUBE *Transverse Brain Divisions*

Everyone knows that the brain and spinal cord are inside the body, not on the surface. How does the neural plate end up forming the internalized central nervous system? The answer is that, in essence, the walls of the neural plate (the neural folds) assume a vertical orientation, the dorsal lips fuse, and the resulting tube sinks into the body, between the dorsal surface ectoderm (future skin) and the notochord (Fig. 4.8). This process of turning the neural plate into the

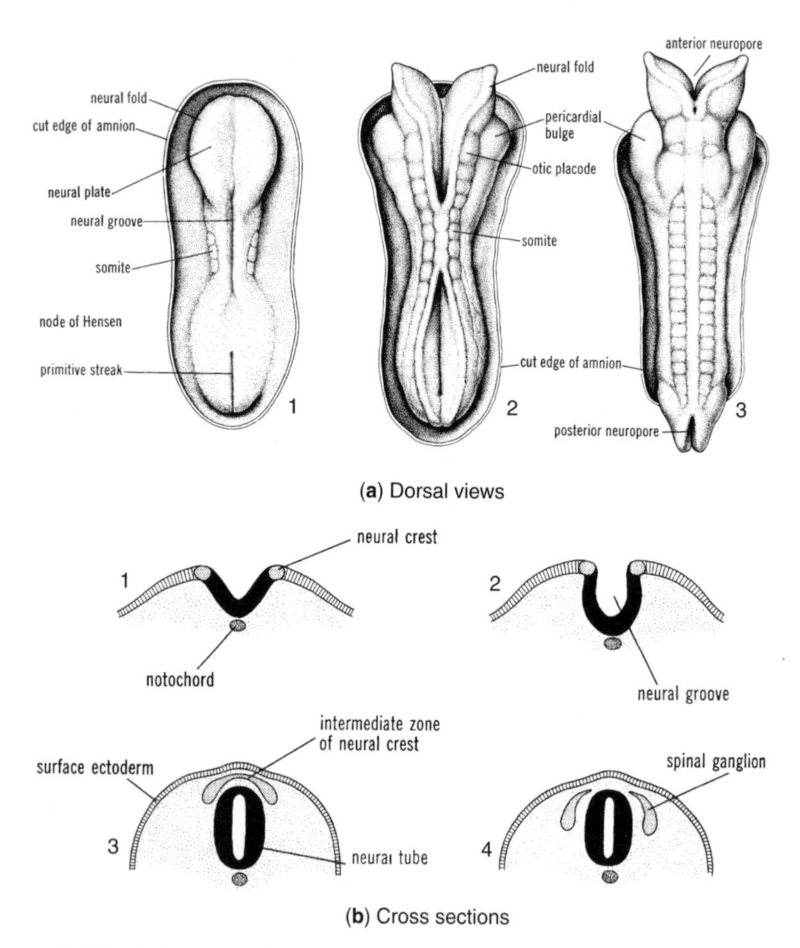

FIGURE 4.8 *Formation of the neural tube in humans is illustrated here from a dorsal perspective (a) and from cross sections (b). Dorsal embryo 1 was 19 days old, embryo 2 was 22 days old, and embryo 3 was 23 days old. Adapted with permission from J. Langman,* Medical Embryology, *third edition (Williams and Wilkins: Baltimore, 1975, pp. 62–63).*

neural tube is called *neurulation.* In mammals, the dorsal fusion tends to start near the transition between brain and spinal cord, basically in the future neck region, and then extends rostrally and caudally until a closed tube is formed. From a topological point of view, it is important to note that the midline of the neural plate becomes the ventral midline of the neural tube, whereas the lateral margins of the neural plate become the dorsal midline of the neural tube.

At the time when the brain region of the neural tube is finally completely closed at the rostral end, it displays three rather distinct swellings or vesicles (Fig. 4.9), just as Malpighi first suggested. Therefore, it should come as no great surprise that at least hints of these three vesicles should be detected later in neural plate growth. Baer provided a great practical service to neuroanatomy when he gave simple, clear names to these rostrocaudally arranged transverse vesicles: forebrain, midbrain, and hindbrain. The earlier nomenclature starting with Malpighi was based on attempts to apply the names of poorly understood adult parts to the early neural tube, and this led to profound confusion. Today, Baer's names and interpretation form the cornerstone of regional or topographic neuroanatomical nomen-

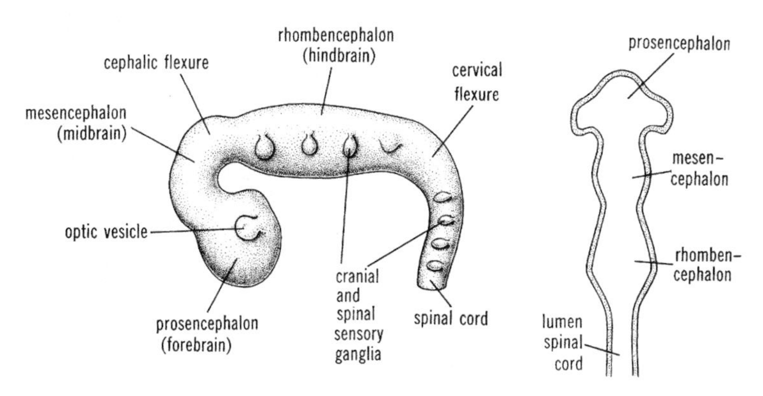

FIGURE 4.9 *The appearance of the three-vesicle stage of the human neural tube (in a 4-week embryo) is shown in a lateral view* (left) *and in a horizontal section through a "straightened out" neural tube* (right). *Reproduced with permission from J. Langman,* Medical Embryology, *third edition (Williams and Wilkins: Baltimore, 1975, p. 320).*

clature, as a glance at the table of contents of most neuroanatomy textbooks published in the last century shows. He suggested that there are three primary brain vesicles, which go on to subdivide into a series of five secondary brain vesicles. In this scheme, the retinal region is in the forebrain vesicle and the otic region is in the hindbrain vesicle.

The five-vesicle stage of the neural tube arises from a subdivision of the forebrain and hindbrain vesicles (Fig. 4.10). When viewed from above or below, the hindbrain vesicle becomes roughly diamond-shaped (hence, *rhombencephalon*), with the rostral half going on to form the pons and the caudal half the medulla (oblongata). Differentiation of the forebrain vesicle is a bit more complex, although in essence a deep groove (when viewed from the outside) forms rostrodorsally to

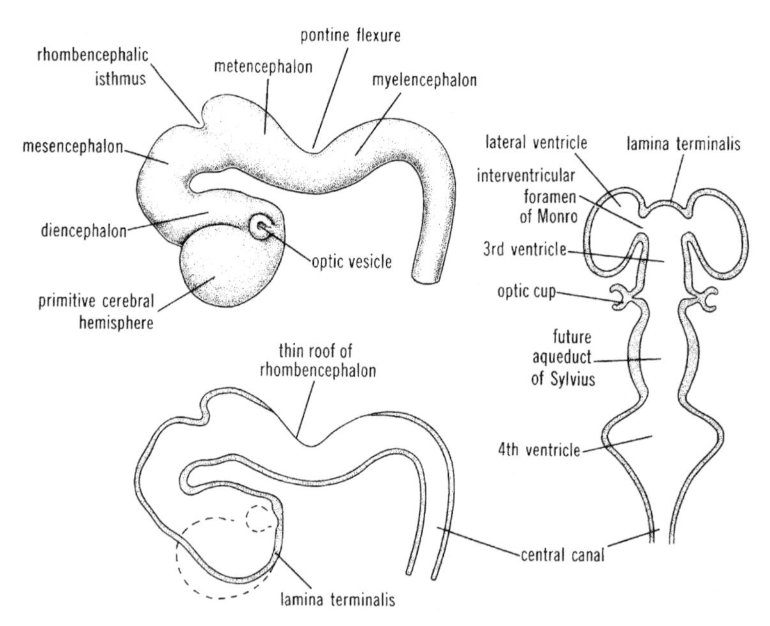

FIGURE 4.10 *The appearance of the five-vesicle stage of the human neural tube at the beginning of the sixth week of development is illustrated here in the same way as the three-vesicle neural tube in the previous figure. Reproduced with permission from J. Langman,* Medical Embryology, *third edition. (Williams and Wilkins: Baltimore, 1975, p. 321).*

produce an endbrain vesicle, followed by an interbrain vesicle. The endbrain (or *telencephalon*) is also called the *cerebral hemisphere*, and there is one on either side—that is, the endbrain vesicles or cerebral hemispheres are paired structures at the rostral end of the neural tube. The interbrain is also known as the *diencephalon*, and the neuroepithelial patch that generates the retina and optic nerves lies within it.

To recapitulate, at the earliest stage of neural tube differentiation, there are three transverse swellings that are arranged from rostral to caudal: the forebrain, midbrain, and hindbrain vesicles. At the next stage, the forebrain and hindbrain swellings divide again so that now we have six rostrocaudally arranged secondary vesicles: endbrain (a pair), interbrain, midbrain, pons, and medulla, which of course is followed by the spinal cord. Amazingly enough, the wall of the early five-vesicle-stage neural tube is still made up entirely of the monolayer neuroepithelium; no neurons have been born yet, although regionalized patterns of gene expression have been reported.

It is obvious that the early neural tube has a segmented appearance, which is even easier to appreciate if the tube is artificially straightened out and sliced horizontally (Figs. 4.9 and 4.10, right side of each). Late-nineteenth-century embryologists referred to these transverse differentiations as *neuromeres*, and it is still not clear whether they are segments (metameres) in the true sense of serial homologous units formed by a common program of gene expression. In any event, they have received a great deal of attention, and it now seems clear that they are proliferation zones for neurogenesis. A series of transitory neuromeres in the hindbrain, the rhombomeres, are especially intriguing and appear to be related to the early differentiation of the cranial nerve nuclei and the adjacent gill slits, another of the core vertebrate features, which are called *pharyngeal* or *branchial arches* in birds and mammals.

NEURAL CREST AND PLACODES *Peripheral Nervous System*

The neural plate represents the central nervous system, with its brain and spinal cord. However, there is a narrow "transition zone" between the neural plate and the somatic ectoderm that pinches off when the neural tube separates from the surface ectoderm and sinks into the interior of the embryonic body (Fig. 4.8). This crest region of ectoderm, the neural crest, goes on to form most of the ganglia of the peripheral nervous system. After pinching off from the ectodermal layer, neural crest cells migrate ventrally into the developing body for greater or lesser distances (Fig. 4.11). As a broad generalization, the neural crest cells that remain closest to the neural tube form the sensory ganglia, those that migrate the farthest end up constituting the massive enteric nervous system in the wall of the gut, and those in between form the autonomic ganglia.

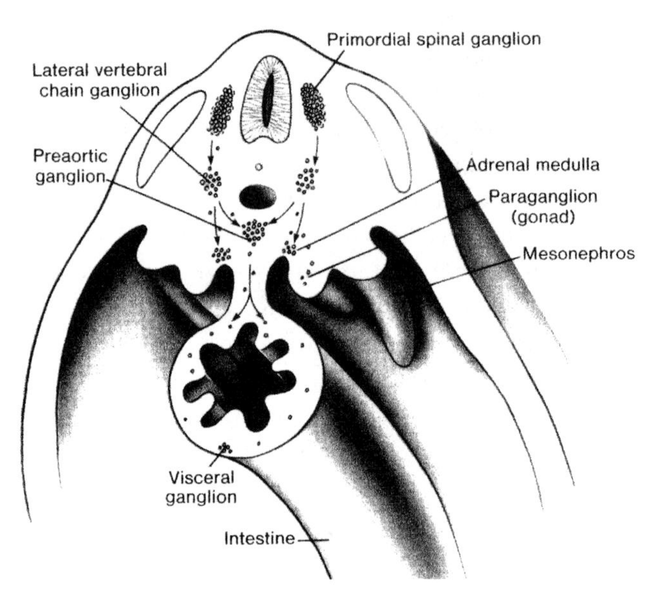

FIGURE 4.11 *The ventral migration of neural crest cells from their early position on either side of the neural tube down into the viscera. As described in the text, neural crest cells form sensory ganglia (primordial spinal ganglia), autonomic ganglia (lateral vertebral chain ganglia, preaortic ganglia, and paraganglia), and the enteric nervous system (visceral ganglia). The neural crest also forms the medulla of the adrenal gland, which is a highly modified autonomic ganglion. Reproduced with permission from M.B. Carpenter, and J. Sutin,* Human Neuroanatomy, *eighth edition (Williams and Wilkins: Baltimore, 1983, p. 69).*

Pairs of sensory ganglia are arranged more or less regularly along the right and left sides of the brainstem and spinal cord. As the name implies, they are collections of sensory neurons that transmit information from various parts of the body (skin, muscle, blood vessels, viscera, and so on) to the central nervous system (Chapter 9). The processes of sensory ganglion cells form important components of the peripheral nerves. In his first publication as a young medical student, Sigmund Freud made the interesting discovery (in 1877) that in the most primitive vertebrates (lamprey) sensory ganglion cells are found within the spinal cord, as well as in sensory ganglia adjacent to the cord. Some 14 years later the great Swedish neuroanatomist Gustav Retzius discovered that in an even more primitive animal, amphioxus (which is in the subphylum Cephalochordata), all sensory "ganglion" cells are found in the spinal cord; we now know that there is even a sensory ganglion in the brainstem of mammals—the mesencephalic nucleus of the trigeminal nerve.

Autonomic ganglia are actually collections of motoneurons that innervate the viscera. Their distribution, and the organization of their connections, is exceptionally complex and poorly understood. Broadly speaking, they fall into two anatomically and functionally distinct subsystems: sympathetic and parasympathetic, which will be discussed in Chapter 6. They are responsible for the largely involuntary or unconscious motor control of the viscera during both sleep and wakefulness, and axons entering and leaving them are fundamental components of the peripheral nerves. Like sensory ganglia, autonomic ganglia are associated with both the spinal cord and the brainstem.

The enteric nervous system is found in the wall of the alimentary tract, where it is concentrated in three concentric, interconnected layers: the outer myenteric plexus of Auerbach, the middle submucosal plexus of Meissner, and the inner mucosal plexus. It is a vast system (with about as many neurons as the spinal cord), and it displays intrinsic activity that is responsible for generating peristaltic waves and many other activities in the alimentary tract. Activity of the enteric nervous system is modulated by inputs from the autonomic nervous system. As mentioned in Chapter 2, the enteric nervous system seems to display many of the features of a complex nerve net.

For the sake of completeness, one other feature of nervous system development needs to be mentioned: the sensory placodes are tiny patches or islands of ectoderm that lie outside the classic neural plate and crest and are specialized to generate sensory neurons. Two types of sensory placodes have been identified. One consists of a series of about five placodes adjacent to the hindbrain segment of the neural crest; these epibranchial placodes either generate or contribute to the sensory ganglia of cranial nerves V (trigeminal), VII (facial, intermediate part), VIII (vestibulocochlear), IX (glossopharyngeal), and X (vagus). The other consists of the olfactory placode, which lies near the prospective endbrain region of the neural plate and generates olfactory sensory neurons (cranial nerve I).

GENERATING NEURONAL CELL TYPES AND GROUPS
Longitudinal Brain Divisions

At the stage of development when the neural tube has differentiated just enough to recognize five brain vesicles and the spinal cord, the wall of the tube is still a simple neuroepithelium. No neurons have been generated. This soon changes, however, and the way in which it happens provides fundamental insights into the basic plan of the central nervous system that Wilhelm His described so beautifully.

Let us begin in the spinal cord, where the pattern is especially clear (Fig. 4.12). When it is viewed in cross section, one can see that the early spinal cord actually has four parts: a thin roof plate in the dorsal midline, a thin floor plate in the ventral midline, and two thicker walls on the right and left sides. At a specific period in development, certain neuroepithelial cells stop dividing and migrate out of the neuroepithelium (called the *ependymal or ventricular layer*) to form a new zone that is referred to as the *mantle layer* of the neural tube. These cells have undergone an irreversible developmental process called *determination* and will never divide again; they are young neurons. Note that the mantle layer of young neurons is sandwiched be-

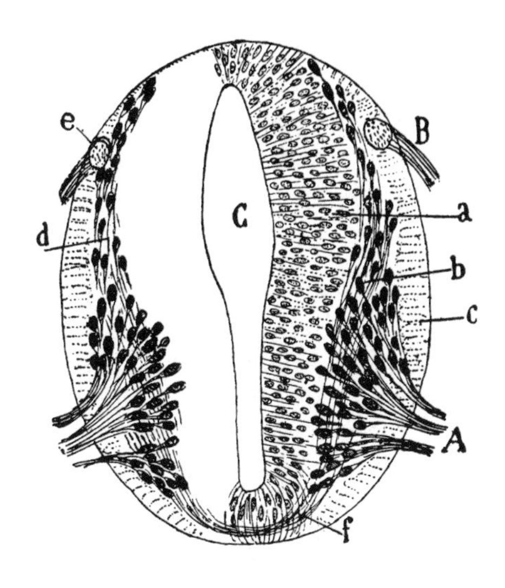

FIGURE 4.12 *In spinal and hindbrain regions, the wall of the neural tube shows a gradient of differentiation from ventral to dorsal. Neurons are born first ventrally, and they go on to become motor neurons and send their axons out of the tube to form the ventral roots (A). It is easy to see in this cross-section of the early spinal cord that the ependymal or ventricular layer (a) is thicker more dorsally, whereas, conversely, the mantle layer of young neurons (b, d) is thicker more ventrally. The neuroepithelium is divided into two separate halves by two midline structures, floor plate (f) and roof plate.* Key: *B and e, dorsal root fibers entering the marginal zone of the neural tube. From S.R. Cajal,* Histologie du système nerveux de l'homme et des vertébrés, *vol. 1 (Maloine: Paris, 1909).*

tween two other layers. The inner is called the *ependymal layer* (which has a pseudostratified appearance and lines the fluid-filled center of the neural tube, the future ventricular system of the adult), and the outer relatively cell-free zone is called the *marginal layer*, which contains the processes of various cell types.

But the deceptively simple drawing in Figure 4.12 reveals much more. On closer inspection it is obvious that the mantle layer at this early stage of development is thicker ventrally than dorsally. In fact, neurogenesis begins ventrally and gradually spreads dorsally—there is a ventral to dorsal gradient of neuron generation in the spinal cord (and, as we shall see, in the hindbrain as well). As a result, the ependymal layer is thinner, and the mantle layer is thicker, ventrally. This arrangement causes a shallow groove to appear on the inner wall of the neural tube. Wilhelm His named this longitudinal groove the *limiting sulcus* and pointed out that it roughly divides each wall of the early neural tube into a ventral basal plate and a dorsal alar plate.

The fundamental significance of the early basal and alar plates was immediately obvious to His. The first neurons to be generated in the neural tube are motor neurons, and their axons grow out of the neural tube in bundles called *ventral roots*. In contrast, the axons of sensory neurons in the dorsal root ganglia grow into the alar plate, whose neurons do not extend axons into the ventral roots. Thus, the early basal plate is associated with the motor system, whereas the early alar plate is associated with the sensory system.

This clear embryological distinction between basal plate / ventral root and alar plate / dorsal root, and its obvious association with distinct functional systems—motor and sensory, respectively—was a brilliant confirmation of what has been called the greatest single discovery in the history of neuroscience, the Bell-Magendie law. Charles Sherrington considered it second only to William Harvey's discovery of the circulation of the blood in the history of physiology. To make a long and regrettable story short, François Magendie published elegant experimental proof in 1822 that the dorsal roots transmit sensory information whereas the ventral roots transmit motor information, thus shattering ancient beliefs that sensory and motor information are transmitted by the same fibers. In essence, this suggested a "circulation" of neural information into the spinal cord via the dorsal roots and out of the spinal cord via the ventral roots. This was a fundamental part of the thinking that went into the gradual development of modern concepts of the reflex arc (Chapter 5) because sensory and motor functions could be distinguished unequivocally on both anatomical and physiological grounds. In the years following 1822, Sir Charles Bell reprinted some of his earlier papers, correspondence, and a private pamphlet and very selectively altered their contents so that he could claim priority for Magendie's discovery.

Now, returning to the central nervous system. As its differentiation progresses, various regions of the neuroepithelium (the ependymal layer) generate different neuronal cell types in a highly stereotyped spatiotemporal pattern, and these young neurons migrate out into the mantle layer along more or less direct radial or tangential routes before settling down to establish connections. Thus, as embryogenesis progresses, the wall of the neural tube becomes thicker and thicker, and each of the brain vesicles becomes characteristically differentiated in very complex ways. Nevertheless, even in fully mature adults, the vertebrate central nervous system is, from a topological perspective, nothing more than a closed tube with highly differentiated walls. And it tends to maintain the three concentric layers found in the embryonic neural tube: an ependymal monolayer lining the central fluid-filled cavity (the ventricular system), a very thick mantle layer of neuronal cell bodies, and a thinner marginal layer of cell processes. This arrangement is crystal clear in the adult spinal cord (Fig. 4.13) and more or less evident in most parts of the brain.

In a very general way, the walls of the adult central nervous system can be divided into what are loosely referred to as gray matter

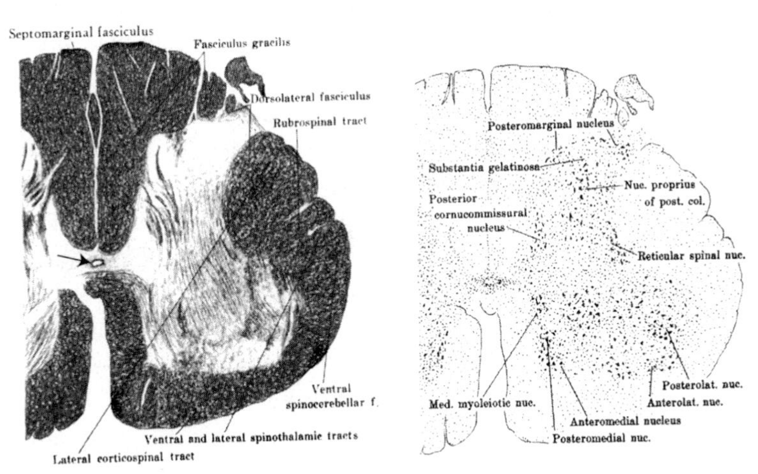

FIGURE 4.13 *These photomicrographs of histological sections illustrate the arrangement of fiber tracts* (left) *and neuronal cell bodies* (right) *in cross-sections of the adult human spinal cord. The arrow in the figure on the left indicates the central canal, which is the remnant of the neural tube lumen. Its wall is formed by the one-cell thick ependymal layer, all that is left of the ependymal or ventricular layer of the neural tube. Compare with the previous figure, of the embryonic spinal cord. The left example is stained with the Weigert method; the right with the Nissl method (see Appendix C). Reproduced with permission from F.A. Mettler,* Neuroanatomy, *second edition. (Mosby: St. Louis, 1948, p. 226).*

and white matter because of their appearance to the naked eye when one slices into the brain or spinal cord. White matter consists of major fiber tracts—that is, the major collections of axons that course longitudinally or transversely throughout the central nervous system. In contrast, gray matter is characterized by the presence of huge numbers of neuronal cell bodies that are not distributed uniformly but are aggregated into more or less identifiable collections referred to as *cell groups* (which give rise to the fiber tracts). In an oversimplified way, fiber tracts are analogous to the highway system on a map, whereas cell groups are analogous to the cities and towns where the highways begin, end, or pass through.

When various specialized stains are applied to thin sections of central nervous system tissue (Appendix C), the cell groups become clear (Fig. 4.13, right). It must be admitted, however, that it is easy to see a border around some of them and not so easy (or even impossible) to see a border around others that seem to merge imperceptibly into one another. One important reason that cell groups can even be distinguished is that there are many different neuronal cell types (Chapter 2), which can vary in size, shape, staining intensity, and packing density. Fortunately, particular cell types tend to cluster in recognizable cell groups. Because of this, different cell groups have different functions, and a catalog of major cell groups amounts to a "parts list" of the central nervous system (Appendix B).

The only reason cell groups can be recognized at all is that they show some pattern of cell staining that distinguishes them from other cell groups. However, the structure of cell groups is more complicated than this: it is the rule rather than the exception that cell groups are formed by more than one interdigitated cell type. Furthermore, it is not uncommon that the interdigitated *cell types are distributed in complex gradients* that are particular for each cell type. As a matter of fact, each cell group in the central nervous system probably has a unique cellular architecture that cannot be determined in anything other than an empirical way. This particular aspect of neuroanatomy is referred to as *cytoarchitecture.*

It cannot be overemphasized that, in the long run, the circuitry of the brain is best described in terms of cell types and not cell groups.

This is not a straightforward situation, however, because there are good examples of clear cell types that are not restricted to a particular cell group or layer. For example, all retinal ganglion cells are not in the ganglion cell layer; there are displaced retinal ganglion cells. The rigorous analysis of brain circuit architecture must be based on a description of neuronal cell type distribution patterns with reference to the regionalization map of cell groups or basic parts (see Appendix A).

For convenience, cell groups are often divided into two broad categories: laminated and nonlaminated. As the name implies, laminated cell groups display layers, and if they lie on the surface of the brain, they are often referred to as *cortex* (although by tradition, "cortex" has been reserved for laminated surface regions of the cerebral and cerebellar hemispheres). In contrast, nonlaminated cell groups are usually referred to as *nuclei* (the term was first used in this way by the neuroanatomist Johann Christian Reil in 1809), although nonlaminated cell groups with relatively indistinct borders are often referred to as *areas* or *regions* instead of nuclei. For clarity and consistency, there is a now a strong preference to restrict in vertebrates the use of the term *ganglion* to distinct collections of neurons in the peripheral nervous system. The problem is that, historically, any distinct group of neurons in the central or peripheral nervous system was referred to as a *ganglion*, and this usage has lingered in some current versions of neuroanatomical nomenclature. For example, the nonlaminated mass of the cerebral hemisphere is referred to as the *basal ganglia* in some textbooks and as the *basal nuclei* in others.

There is no fundamental reason why one cell group is laminated and another is nonlaminated. As a matter of fact, an homologous cell group may be laminated in one species and nonlaminated in another. For example, the lateral geniculate nucleus, which relays sensory information from the retina to primary visual cortex in the cerebral hemisphere, is distinctly laminated in cats but not in rats. The primary taste relay cell group in the hindbrain, the nucleus of the solitary tract, presents an especially curious and dramatic case. In most fish, as in most other vertebrates, it is a nonlaminated cell group along the dorsomedial surface of the lower hindbrain, near the beginning of the spinal cord. However, in certain fish, the "nucleus" of the solitary tract forms a huge laminated vagal lobe on either side of the brainstem (Fig. 4.14). In these fish, taste buds have spread from the mouth and tongue to cover the surface of the fish in a huge "gustatory map" that seems to be reflected in the structure of its sensory relay cell group in the hindbrain. It would appear that the architecture of a cell group (as well as the shape of individual neurons) is dramatically influenced by the organization of its neural inputs, which is established during embryogenesis.

For the sake of completeness, I should mention that fiber tracts can range from simple to exceptionally complex, and from well circumscribed to diffuse and indistinct. At one extreme, we could cite the axons from the trochlear nucleus in the brainstem. The trochlear nucleus contains motoneurons that innervate just one of the muscles that move the eyeball, and their axons course together through the brainstem in a very compact bundle or root until they leave to form the trochlear nerve outside the brain. The trochlear nerve root is a very discrete, simple fiber tract in the brain. At the other extreme, the medial forebrain bundle has on the order of a hundred intermixed components, and it is diffusely organized with no clear borders anywhere. At the same time, it is exceptionally important functionally: in essence, it is responsible for the expression of motivated and emotional behaviors.

FIGURE 4.14 *The appearance of the vagal lobe sensory region (nucleus of the solitary tract) in two species of fish is illustrated in these Nissl-stained transverse sections through the medulla (lower brainstem). In the catfish (a), this sensory input region of the vagus nerve has the typical appearance of a nucleus; in the goldfish (b), in contrast, it has a highly elaborate laminated appearance. Note how massive the vagus nerve is in the goldfish. One sensory modality transmitted by the vagus nerve is gustation, and the goldfish has an incredibly differentiated gustatory system. They have a high density of taste buds distributed throughout the lining of the oropharyngeal cavity, including the surface of the branchial (pharyngeal) arches. Inputs from these taste buds are mapped onto a highly differentiated, laminated vagal lobe. The gustatory system in catfish is much less differentiated, and this is reflected in the "nuclear" organization of its vagal lobe. Photomicrographs kindly provided by Thomas E. Finger.*

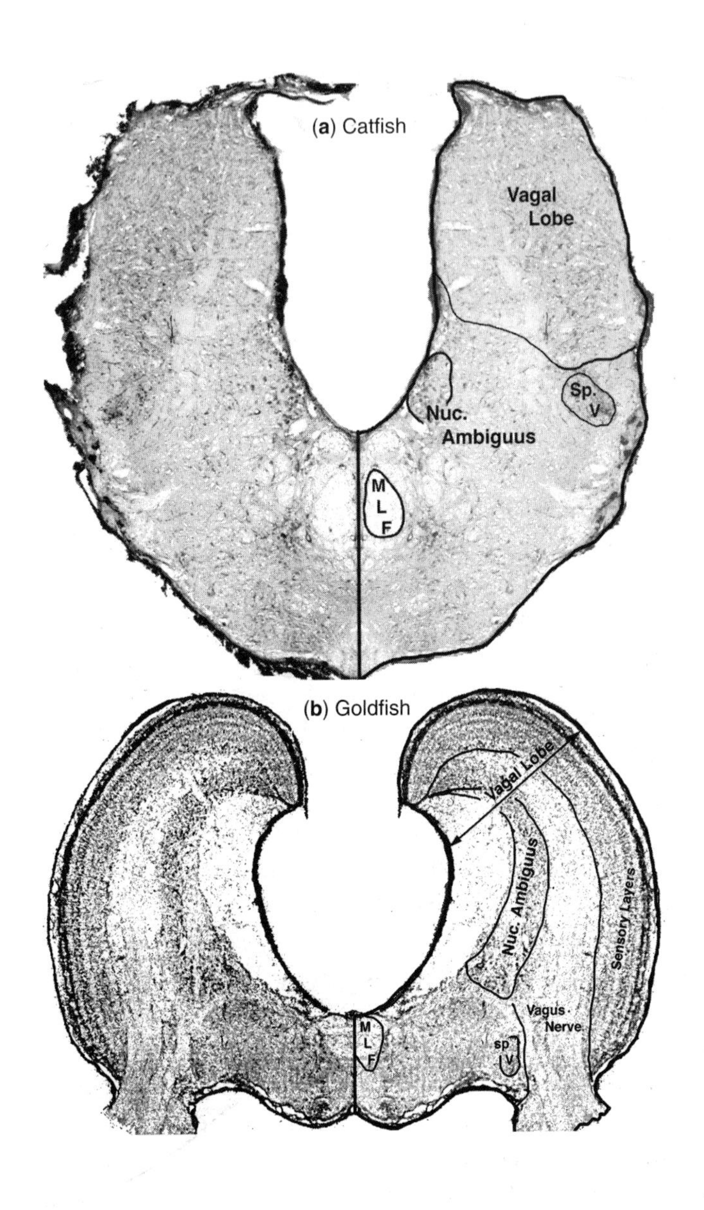
(a) Catfish
Vagal Lobe
Sp. V
Nuc. Ambiguus
M L F
(b) Goldfish
Vagal Lobe
Nuc. Ambiguus
Sensory Layers
Vagus Nerve
M L F
sp V

A NERVOUS SYSTEM FATE MAP

We have just learned that motor neurons tend to develop first, in the ventral or basal plate of the neural tube, ventral to the limiting sulcus—after the primitive neural tube has differentiated into five brain vesicles and the spinal cord. In principle, we could take a pair of scissors, cut along the dorsal midline of the neural tube from the caudal to the rostral end, and then flatten it out, which, in essence, would return it to the neural plate stage. In this imaginary flatmap, the ventral midline of the early neural tube would form the midline and the dorsal midline would form the lateral borders, just as we have seen in the neural plate before it forms the neural tube (see Fig. 4.8). But, most important, this neural tube flatmap would illustrate with crystal clarity the basic topological principles of central nervous system development discussed earlier in this chapter: transverse differentiation into rostrocaudally arranged endbrain, interbrain, midbrain, pons, medulla, and spinal cord; and longitudinal differentiation into ventral (basal) and dorsal (alar) plates.

Having taken the conceptual leap of flattening the early neural tube, let us go ahead and ask how that flatmap is related to the neural plate itself. The most obvious approach is to assume that specific parts of the neural tube are generated by specific parts of the neural plate. As a simple example, the widened, rostral half of the early neural plate forms the brain, whereas the narrower, caudal half of the neural plate forms the spinal cord. It does not take a great deal of imagination to hypothesize further that the rostral end of the brain plate generates the forebrain, whereas the caudal end of the brain plate generates the hindbrain, and so on. Indeed, this line of reasoning was taken rather far by Wilhelm His in the nineteenth century (see Fig. 4.7), and it has been pursued experimentally by a number of investigators since then. These researchers have developed fate maps of the neural plate: they are prospective regions of the neural plate that will go on later to generate specific cell groups in the neural tube. Experimental fate mapping has led to a broad understanding of how the major divisions of the central nervous system are represented in the neural plate and early neural tube, although a great deal remains to be learned about the finer subdivisions, especially with the analysis of relevant gene expression patterns.

Figure 4.15 is one version of a fate map of the mammalian neural plate (central nervous system), and if nothing else it is a useful visual aid for describing how the neural tube differentiates as neurogenesis progresses. The left side of the fate map shows the approximate prospective locations of the forebrain, midbrain, hindbrain, and spinal cord, which are basically transverse blocks arranged from rostral to caudal (compare with the three-vesicle stage neural tube in Fig. 4.9). In contrast, the right side of the fate map shows the next major stage of neural development, the five-vesicle stage. At this stage the brain has become divided into endbrain and interbrain vesicles by the hemispheric sulcus, and the hindbrain has a rostral pons and a caudal medulla (compare with Fig. 4.10). There seems to be almost universal agreement about this basic transverse structural organization of the central nervous system.

There is less agreement about the longitudinal organization of the central nervous system, and controversy about this grows exponentially as one goes rostrally into the midbrain and then forebrain. Everyone seems to agree that the limiting sulcus of the early neural tube (see the preceding section) can be traced uninterrupted from the caudal tip of the spinal cord all the way to the rostral end of the hindbrain—that is, to the junction between pons and midbrain. This is interesting in view of B.F. Kingsbury's observation in the 1920s that the histologically defined floor plate of the vertebrate neural tube also stops at the pons/midbrain junction, and it suggests that alar and basal plates, which are so characteristic of the spinal cord, extend uninterrupted through the hindbrain.

In a way, the hindbrain is a rostral extension of the spinal cord that contains motor, sensory, and other cell groups associated with cranial nerves rather than with spinal nerves. However, the hindbrain region of the neural tube has one major feature that distinguishes it absolutely from the spinal cord: a dorsal longitudinal zone known as the *rhombic lip*. This characteristic hindbrain specialization,

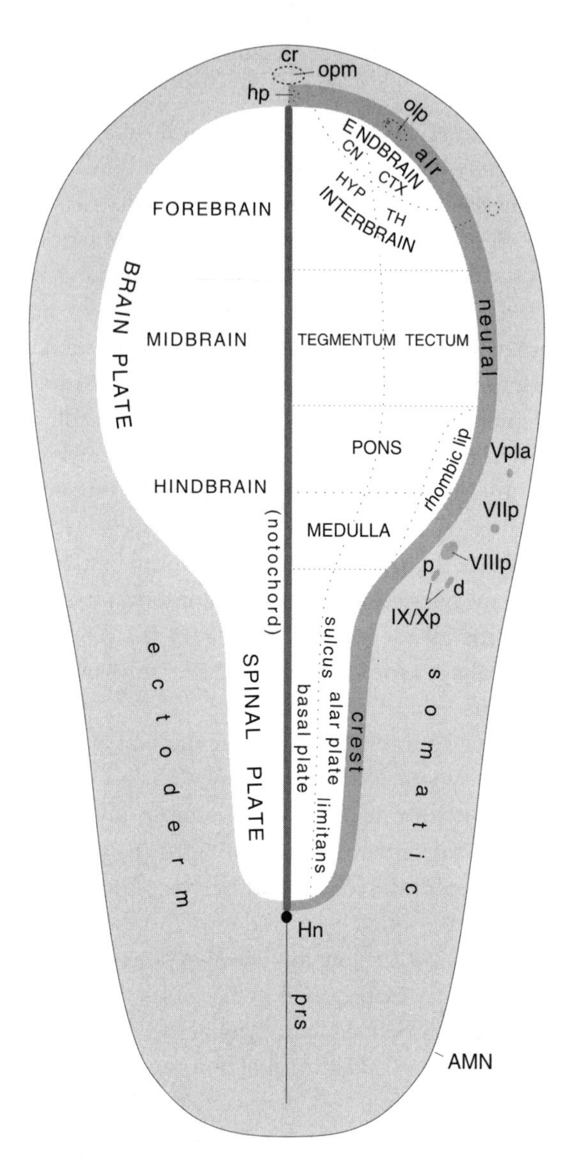

FIGURE 4.15 *This is a fate map of the embryonic ectoderm, including its neural and somatic components.* Key: *alr, anterolateral ridge; AMN, cut edge of amnion; CN, cerebral nuclei (basal nuclei or ganglia); cr, cardiac region; CTX, cerebral cortex; Hn, Hensen's node; hp, hypophysial placode; HYP, hypothalamus; IX, Xp (d, p), glossopharyngeal, vagal placodes (distal, proximal); olp, olfactory placode; opm, oropharyngeal membrane; prs, primitive streak; TH, thalamus; Vpla, trigeminal placode; VIIp, facial placode; VIIIp, otic placode. Reproduced with permission from L.W. Swanson,* Brain Maps: Structure of the Rat Brain (*Elsevier Science: Amsterdam, 1992, p. 25).*

whose presumptive region is found along the lateral margin of the neural plate, generates neural structures that clearly differentiate the hindbrain from the much simpler spinal cord. These structures include special sensory nuclei (for example, associated with hearing, balance, and the viscera), the cerebellum, and certain nuclei associated with the cerebellum (for example, the pontine gray and lateral reticular nucleus). It is almost as if the rhombic lip zone was added on top of the spinal cord architecture (as a matter of fact, it lies on top of the spinal cord extension into the hindbrain, the trigeminal complex; see Chapter 9).

According to most neuroembryologists who have examined the problem carefully, the limiting sulcus cannot be traced uninterrupted into the midbrain vesicle of the neural tube, and this is where uncertainty about the basic longitudinal organization of the central nervous system begins to creep in. What does seem to be true is that there are two longitudinal sulci or grooves running along the early midbrain vesicle. The more dorsal of the two grooves divides the vesicle into a dorsal "tectal" (roof) region and ventral "tegmental" (floor) region, whereas the more ventral sulcus divides the tegmental region into dorsal and ventral zones. The most important generalization about the midbrain is that sensory functions are usually ascribed to the tectum and motor functions are usually ascribed to the tegmentum.

Now we come to the forebrain vesicle, the most complex and uncertain of all. The first thing that happens after the endbrain and interbrain vesicles differentiate within it is the appearance of two longitudinal grooves in the interbrain: the hypothalamic and middle interbrain sulci. They are a consequence of the first neurogenesis in the forebrain vesicle, which takes place in the prospective region of a structure called the *ventral thalamus*, between the thalamus (dorsally) and hypothalamus (ventrally). Remember that in the spinal cord and hindbrain, there is a ventral to dorsal gradient of neurogenesis. In the interbrain, neurogenesis begins instead in an intermediate longitudinal strip (actually an arch that includes the ventral thalamus and retrochiasmatic region of the hypothalamus), then spreads to the bulk of the hypothalamus, and finally begins in the thalamus. And things get even more complicated: a third groove, the habenular sulcus, appears just ventral to the interbrain roof plate. The habenular sulcus divides the thalamus into epithalamus (most dorsal) and dorsal thalamus (between epithalamus and ventral thalamus). As a result, the interbrain vesicle can be divided into four roughly longitudinal strips, arranged from dorsal to ventral: epithalamus, dorsal thalamus, ventral thalamus, and hypothalamus. As a very broad generalization, the dorsal thalamus is basically sensory in function, whereas the rest of the interbrain (mostly ventral to the dorsal thalamus) is basically motor in function.

And finally, there is the endbrain vesicle (also known as the *cerebral hemisphere* or *cerebrum*) at the rostrodorsal end of the neural tube. The first sign of differentiation here is also the appearance of a roughly longitudinal groove due to initial neurogenesis in the ventral half of the vesicle. This groove indicates division of the vesicle into its two basic parts, cortex dorsally and basal nuclei ventrally (where the neurogenesis begins). The "corticobasal" sulcus appears at about the same time as the habenular sulcus in the interbrain, and it is followed shortly afterward by another longitudinal sulcus that further divides the basal nuclear region of the endbrain vesicle into a dorsal (striatal) ridge and a ventral (pallidal) ridge. In a very general way, it seems that neurogenesis in the endbrain vesicle progresses from pallidal ridge, to striatal ridge, to cortex. In the adult, it is common to regard the topologically dorsal cortex as having a "sensory" function and the topologically ventral basal nuclei as having a "motor" function.

When we look at the embryonic central nervous system at a time when neurons are just beginning to be generated, we can see in each of the brain vesicles and spinal cord that these early neurons tend to come from ventral regions of the neuroepithelium and tend to have motor functions later on. This is clear in the hindbrain and spinal cord, where a continuous limiting sulcus divides the walls of the neural tube into basal (ventral) and alar plates. We can also see it in the midbrain, where the tectal sulcus divides the walls of the neural tube into tegmental (ventral) and tectal regions; and in the interbrain, where the middle interbrain sulcus divides the walls of the neural tube into the ventral thalamic and hypothalamic region (ventral) and the rest of the thalamus. And finally, we can see it in the endbrain, where the corticobasal sulcus divides the wall of the neural tube into basal nuclear (ventral) and cortical regions. There is no way of knowing at this time whether the tectal, the middle interbrain and the corticobasal sulci are discontinuous, rostral components of an extended limiting sulcus or whether they are completely independent features of the midbrain, interbrain, and endbrain vesicles.

In any event, this seems to be the basic transverse and longitudinal organization of the central nervous system. As embryonic differentiation continues, it is thought that each of the regions defined by the longitudinal and transverse grooves is further subdivided over and over into the final adult complement of cell groups, which of course are interconnected in very specific ways by a variety of fiber tracts. Exactly what the true regionalization plan or fate map of the neural plate actually is remains to be determined, most probably by understanding the genetic program that builds the central nervous system over the course of development.

In the meantime, at least four different schemes, based simply on interpretations of morphology, have been proposed. The original plan advanced by Wilhelm His in the late nineteenth century has already been described in the section on the neural plate (see Fig. 4.7), and it is shown schematically in Figure 4.16. The key features of this interpretation are that *(a)* the presumptive floor plate extends to the rostral end of the neural plate, *(b)* at the earliest stages the rostral end of the neural plate is marked by the presumptive infundibulum—the stalk of the pituitary gland—, and *(c)* the presumptive limiting sulcus, and thus the presumptive basal and alar plates, extend the length of the neural plate. The scheme outlined here is rather similar except that there is no presumptive floor plate (or a very different floor plate), presumptive limiting sulcus, or presumptive basal and alar plates in the midbrain and forebrain (Fig. 4.16, Alvarez-Bolado/Swanson scheme, parts a and b). Nevertheless, if the tectal,

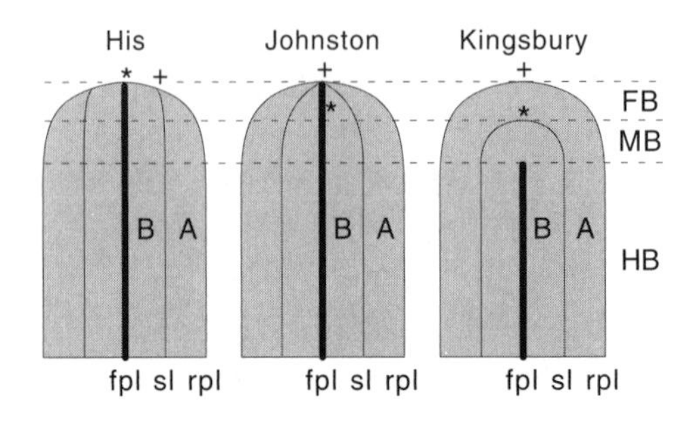

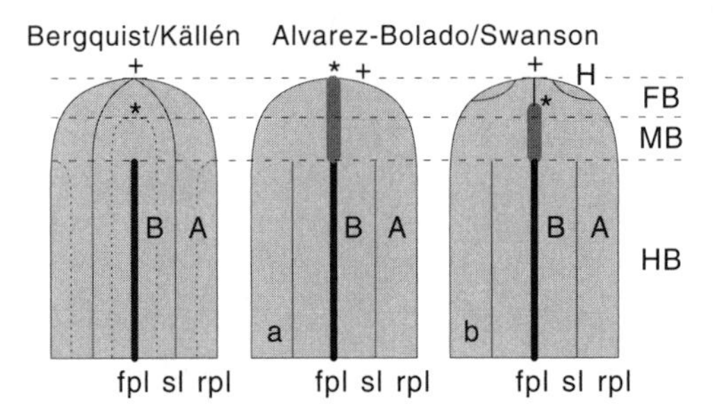

FIGURE 4.16 *Different schemes for neural plate regionalization are illustrated here. The rostral end of the neural plate is shown flattened, in a highly schematic way, with the asterisks (*) indicating the site of the presumptive infundibulum, and the plus-signs (+) indicating the site of the presumptive optic vesicles and chiasm. In the Alvarez-Bolado/Swanson scheme, a and b refer to earlier and later stages, respectively, and the thick lighter gray region above the floor plate is the prechordal plate.* Key: *A, alar plate; B, basal plate; FB, forebrain; fpl, floor plate; H, cerebral hemisphere (endbrain); HB, hindbrain; MB, midbrain; rpl, roof plate; sl, limiting sulcus. Adapted with permission from G. Alvarez-Bolado and L.W. Swanson,* Developmental Brain Maps: Structure of the Embryonic Rat Brain *(Elsevier Science: Amsterdam, 1996, p. 36).*

middle interbrain, and corticobasal sulci turn out to be disconnected rostral extensions of the limiting sulcus, then the two models are remarkably similar. In the early 1920s, Kingsbury proposed a third model where *(a)* the floor plate stops at the hindbrain–midbrain junction; *(b)* the basal plate extends across the midline rostral to the floor plate, and is thus continuous, with an inverted U-shape; *(c)* the alar plate also crosses the midline, rostral to the basal plate, so it too is continuous, with an inverted U-shape; and *(d)* the rostral end of the basal plate is found somewhere near the infundibulum. Finally, yet another scheme was proposed by the obscure yet brilliant Swedish neuroembryologists H. Bergquist and B. Källén in the 1950s. They suggested that the prospective limiting sulcus, along with the prospective basal and alar plates, meet at the rostral tip of the neural plate (Fig. 4.16).

Only time will tell which if any of these basic architectural plans is correct, or even whether this is a valid way of dissecting or parceling the central nervous system. But whatever the case may be, it is nevertheless valid to ask what relationship there is between the transverse and longitudinal parceling of the neural tube, and the organization of functional systems in the young, then mature, and finally aging brain. As argued in the rest of this book, the embryonic approach just outlined is like describing the body in terms of parts such as the head, hands, and feet. In contrast, the functional approach is like describing the body in terms of traditional systems—nervous, muscular, circulatory, and so on. When we think of behavior, it is usually in terms of a particular act such as the hand (a part) reaching for an object. The biological explanation of this behavior must be framed in terms of how all of the functional systems interact over both the short term and the long term. For example, the nervous system controls the musculoskeletal system of the fingers and modulates the blood supply to this active tissue, and so on, and so forth. Parts and systems are ways that biologists have come to describe how the body works; in a logic that I don't fully understand, they are complementary ways of dealing with the same object, the body.

To finish this section, let us return to the neural plate fate map and simply point out the obvious: there is a continuous differentiation of cell groups and fiber tracts in the walls of the neural tube throughout the embryonic period of development. These various structures mature with strikingly different spatiotemporal patterns, and two of the larger units—the cortex of the cerebral and cerebellar hemispheres—differentiate incredibly massively and quite late (even partly after birth) in mammals. As a result, the areal proportions of regions shown on the schematic neural plate fate map in Figure 4.15 (which is based on a very early stage of development) are useless for a flatmap of the adult central nervous system. One way to solve this problem is simply to make the area of a particular cell group in the adult flatmap proportional to its actual weight in the brain—while, of course, preserving boundary relationships between parts as much as possible. The results of this type of transformation for the adult rat central nervous system are shown in Figure 4.17.

OVERVIEW *Parts of the Nervous System*

If this chapter has seemed like a thinly veiled geography lesson, that is exactly what it is—the basic geography of the nervous system. Its like taking a globe and starting with an outline of the major oceans and continents (with their names), and then going on to show in more detail how the continents are divided into countries, the countries into states or provinces, and so on and so forth. It is true that from an historical perspective these boundaries and names are subject to change, but they nevertheless have two exceptionally important functions. First, they provide a vocabulary for describing locations on the surface of the earth. And second, they are used for constructing maps of the earth's surface that are complete, systematic, and geometric (topologic) inventories of geographic places or "parts." As common experience teaches, maps are very handy and useful ways of transmitting geographic information, in a very abstract but at the same time accurate way. And as Gerardus Mercator made crystal clear in the sixteenth century, flatmaps are much handier than globes.

The map we have outlined in this chapter illustrates the basic structural plan of the vertebrate central nervous system based on what little is known about the development of the neural plate, the earliest and simplest representation of the central nervous system in the embryo. We can now go on in the next chapter to ask a fundamental question: what is the basic wiring diagram of the nervous system? This problem needs to be discussed in terms of the various parts of the nervous system outlined here, as well as in terms of how in-

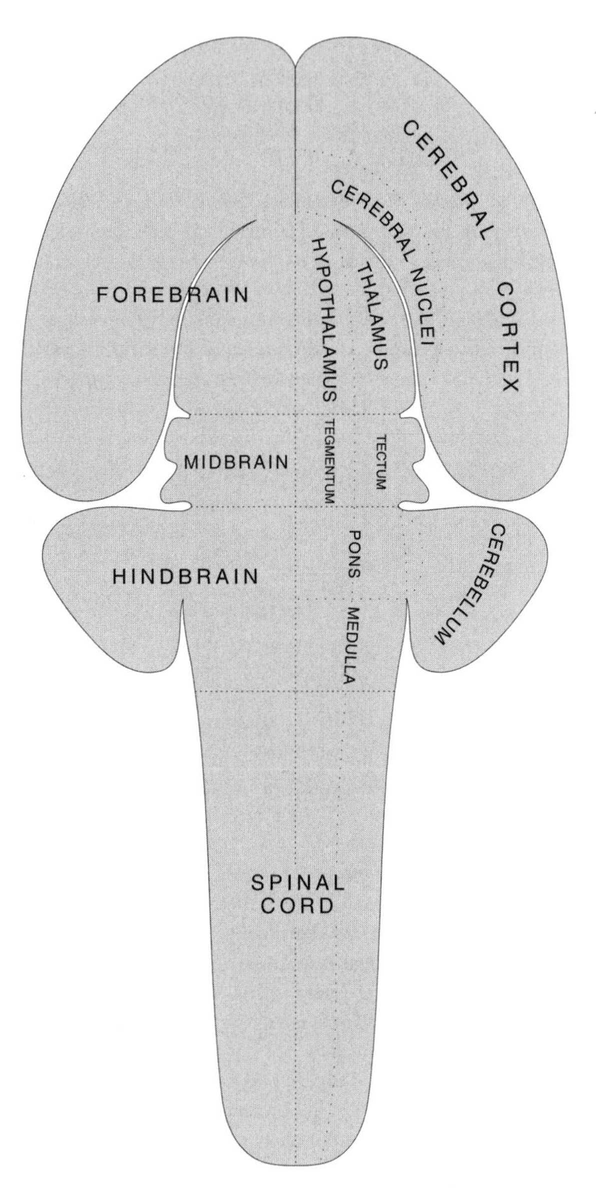

FIGURE 4.17 *This flatmap shows the basic arrangement of the major parts or regions of the adult central nervous system, scaled for the rat. Adapted with permission from L.W. Swanson,* Brain Maps: Structure of the Rat Brain *(Elsevier Science: Amsterdam, 1992, p. 35).*

dividual neurons are interconnected to form specific circuits and networks, using the general concepts developed in Chapters 2 and 3.

READINGS FOR CHAPTER 4

Adelmann, H.B. *Marcello Malpighi and the Evolution of Embryology*, 5 vols. Cornell University Press: Ithaca, 1966. This is an excellent scholarly history of embryology, with a focus on one of its greatest heroes.

Alvarez-Bolado, G., and Swanson, L.W. *Developmental Brain Maps: Structure of the Embryonic Rat Brain.* Elsevier: Amsterdam, 1996. Overview of the literature on rodent brain development with atlases of the various stages.

Arendt, D., and Nübler-Jung, K. Comparison of early nerve cord development in insects and vertebrates. *Development* 126:2309–2325, 1999.

Barteczko, B., and Jacob, M. Comparative study of shape, course, and disintegration of the rostral notochord in some vertebrates, especially humans. *Anat. Embryol.* 200:345–366, 1999.

Bergquist, H., and Källén, B. Notes on the early histogenesis and morphogenesis of the central nervous system in vertebrates. *J. Comp. Neurol.* 100:627–659, 1954.

Cranefield, P.F. *The Way In and the Way Out: François Magendie, Charles Bell and the Roots of the Spinal Nerves.* Futura Publishing: Mount Kisco, N.Y., 1974. A fascinating analysis of scientific misconduct surrounding one of the greatest discoveries in physiology.

Hamilton, W.J., and Mossman, H.W. *Human Embryology: Prenatal Development of Form and Function*, fourth edition. Williams and Wilkins: Baltimore, 1972. The best written and illustrated exposition of classical principles and results in vertebrates.

His, W. *Die Entwicklung des menschlichen Gehirns wäehrend der ersten Monate.* S. Hirzel, Leipzig, 1904. This is the master's final summary, beautifully illustrated.

Holland, P.W.H., and Graham, A. Evolution of regional identity in the vertebrate nervous system. *Persp. Dev. Neurobiol.* 3:17–27, 1995. There are not many answers yet, only some tantalizing hints.

Jacobson, M. *Developmental Neurobiology*, third edition. Plenum Press: New York, 1991. This is an authoritative synthesis of the literature from an historical and almost philosophical vantage.

Keyser, A. The development of the diencephalon of the Chinese hamster. *Acta Anatomica* 83, Suppl. 59, 1972. An exceptionally thorough and insightful review of the literature on the topology of morphological divisions in the mammalian neural tube is presented.

Kingsbury, B.F. The fundamental plan of the vertebrate brain. *J. Comp. Neurol.* 34:461–491, 1922.

Kuhlenbeck, H. *The Central Nervous System of Vertebrates:* Vol. 3, *Part II: Overall Mor-*

phologic Pattern. S. Karger: Basel, 1973. A masterful review of a vast literature on morphological features of the vertebrate neural plate and tube from the topological perspective is provided.

Langman, J. *Medical Embryology: Human Development—Normal and Abnormal,* fourth edition. Williams and Wilkins: Baltimore, 1981. Here is a very good starting point for understanding the basic principles of embryology for the whole animal; renowned for its simple, clear diagrams.

Nieuwenhuys, R., ten Donkelaar, H.J., and Nicholson, C. *The Central Nervous System of Vertebrates,* 3 vols. Springer: Berlin, 1998. This is the most recent comprehensive review of a vast literature.

Patten, I., and Placzek, M. The role of Sonic hedgehog in neural tube patterning. *Cell. Mol. Life Sci.* 57:1695–1708, 2000.

Swanson, L.W. Mapping the human brain: past, present, and future. *Trends Neurosci.* 18:471–474, 1995.

Swanson, L.W. *Brain Maps: Structure of the Rat Brain—A Laboratory Guide with Printed and Electronic Templates for Data, Models and Schematics,* second edition, with double CD-ROM. Elsevier: Amsterdam, 1998–1999. An atlas of the adult rat brain; a complete flatmap of central nervous system cell groups and fiber tracts is presented.

Swanson, L.W. What is the brain? *Trends Neurosci.* 23:519–527, 2000. This is a short history of how the major parts of the brain have been named.

Trainor, P.A., and Krumlauf, R. Patterning the cranial neural crest: hindbrain segmentation and *Hox* gene plasticity. *Nat. Rev. Neurosci.* 1:116–124, 2000.

Williams, P.L. (ed.) *Gray's Anatomy,* 38th (British) edition. Churchill Livingstone: New York, 1995. This is still the bible of anatomy; an invaluable reference for development and adult structure of the body.

Young, J.Z. *The Life of Vertebrates,* third edition. Oxford University Press: Oxford, 1981. This is a classic overview; very readable.

Zhu, Q., Runko, E., Imondi, R., Milligan, T., Kapitula, D., and Kaprielian, Z. New cell surface marker of the rat floor plate and notochord. *Dev. Dynamics* 211:314–326, 1998.

5

Brain and Behavior

A Four Systems Network Model

He who loves practice without theory is like a seafarer who boards a ship without wheel or compass and knows not whither he travels.
—LEONARDO DA VINCI

. . . Since Darwin and Poincaré, Einstein and de Broglie . . . scientific theories play a role in scientific progress that is just as essential as discoveries and the verification of experiments.
—JACQUES ROGER (1997)

General theories about relationships between the nervous system and behavior have a long, and often amusing, history that goes back well before the discovery of electrical impulses and neurotransmitters. Perhaps the first theory of any real merit was elaborated by Plato in his cosmology, *Timaeus.* He divided the mental and behavioral faculties (referred to as the *soul*) into three categories, each associated with a different level of the central nervous system and corresponding level of the body. The divine part concerned with intellect, reason, sensation, and voluntary movement was placed highest, in the brain, within the head. The mortal part dealing with the emotions came next, in upper regions of the spinal cord associated with the thorax, especially the heart. And the baser part subserving the appetites was lowest, in regions of the spinal cord associated with the abdomen and pelvis. Because it is desirable that these functions be partly shielded from one another, the neck forms an isthmus separating the intellect from the emotions, and the diaphragm separates the emotions from the appetites. Furthermore, this was a hierarchically organized functional model: the intellect influences the emotions, which in turn influence the appetites.

The next generation of theories was stimulated about five centuries later by Galen, and it was not completely abandoned for an astonishing 1500 years or so. Initially, the theory rested on two pillars. First, there was the three-compartment ventricular system of the brain, whose anatomy was described so thoroughly by Galen. Second, there was the hypothetical substance or force stored in the ventricles and responsible for nervous system functioning—Aristotle's *psychic pneuma* or animal spirits—the vehicle of the soul. By the tenth century these ideas evolved into a dynamic and generally accepted theory somewhat analogous to digestion, supplemented by the incorporation of Aristotle's basic psychological principles. The theory is beautifully illustrated in Figure 5.1. It shows that all of the senses transmit images to the first ventricle (our right and left lateral ventricles, those of the cerebral hemispheres), which thus corresponds to Aristotle's *sensus communis* ("common sense")—the place where inputs from the individual senses are combined to produce images and imagination. These images are then passed on to the second cell (our third ventricle) where they are manipulated by the process of reasoning. Finally, the residual is sent to the third cell (our fourth ventricle) where it is stored as memory.

The last major addition to this theory was provided by René Descartes in the middle of the seventeenth century. He proposed that the flow of psychic pneuma up and down the hollow nerves was controlled by the soul, which he localized to a central position within the brain, in the tiny, unpaired pineal organ (Fig. 5.2). As you can see, the Galenic model was essentially based on hydraulic principles learned from irrigation and plumbing—processing and regulating the flow of psychic pneuma through nerves, instead of water through ditches and pipes. As time went on, psychic pneuma was replaced with "nerve juice or fluid," then with animal electricity, and now with a combination of electrical impulses and neurotransmitter molecules. And analogies with hydraulic systems and clocks were replaced with analogies to machines, then telephone switchboards, and

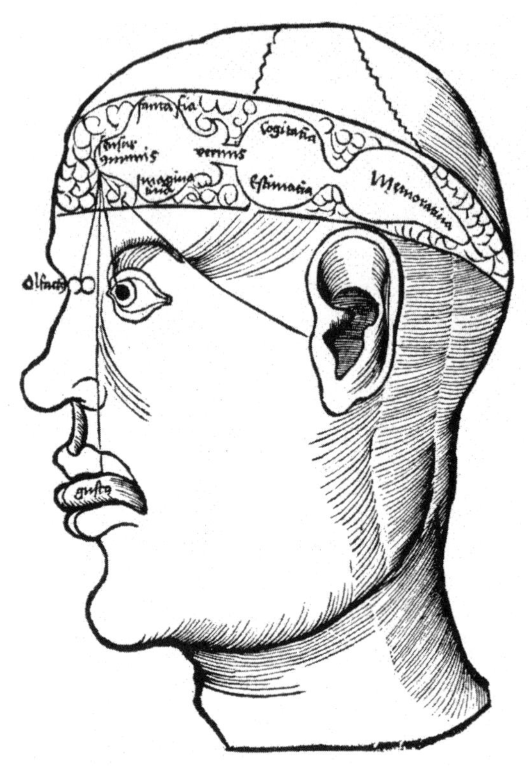

FIGURE 5.1 *Vesalius (1543) mentioned that he used this particular ventricle man drawing in medical school to learn about the brain. It was published by Gregor Reisch in his* Margarita philosophica, *1503, a collection of grammar, science, and philosophy that is considered to be the first modern encyclopedia of any real merit. There is a horizontal "window" through the skull into the brain, showing three interconnected cavities surrounded by curly lines presumably indicating the cerebral convolutions. The rostral end of the rostral cavity or cell (our lateral ventricle) is labeled* sensus communis *(common sense), and this is where all of the special senses (indicated by lines from the sensory organs) converge. More caudally in the first ventricle one finds* fantasia *(fancy) and* imaginativa *(imagination). The passage between the first and second ventricles is labeled* vermis *(worm), and this refers to the choroid plexus "valve" that extends through the interventricular foramen (of Monro). The second ventricle (our third ventricle) is labeled* cogitativa *(thought) and* estimativa *(judgment). The third cell or ventricle in this drawing corresponds to our fourth ventricle, and it is labeled* memorativa *(memory). The passage between the middle and caudal ventricle (which we refer to as the cerebral aqueduct of Sylvius) is unlabeled. For one of Vesalius's renderings of the brain, see Figure 8.10.*

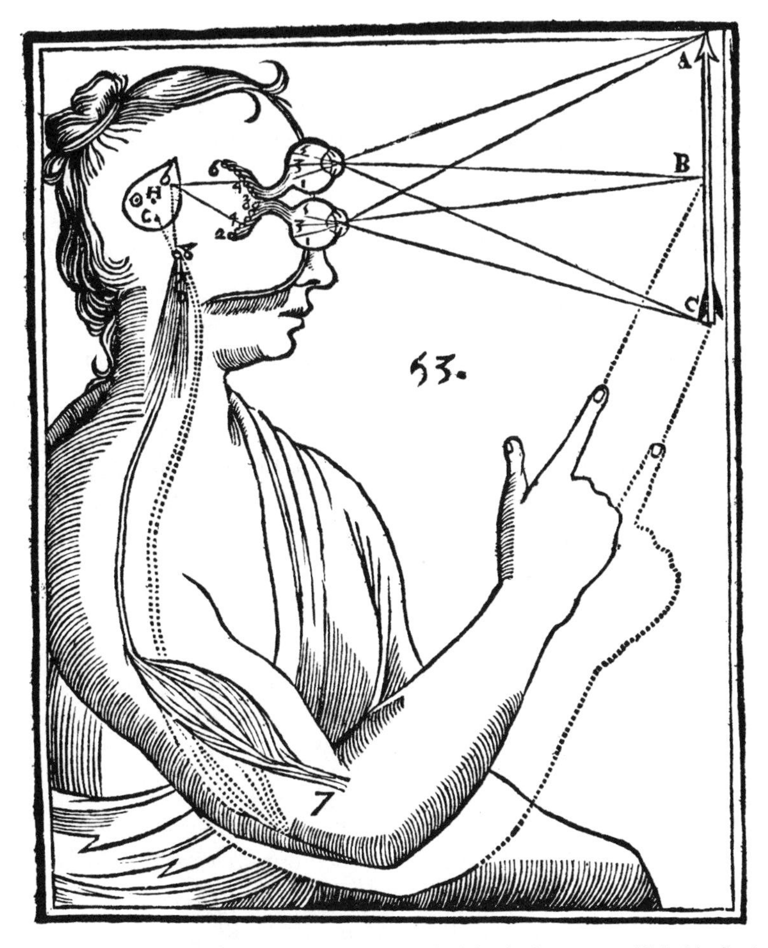

FIGURE 5.2 *The first diagrams illustrating the principle of reflexes were published by René Descartes in his* L'Homme (Treatise of Man). *This illustration is from the 1664 edition in French, which was the language it was written in, and the edition containing illustrations was supervised by Descartes himself. The earliest (unauthorized) edition was published in 1662 in a Latin translation, with very different figures.*

now with computers! Over the centuries, there has been an obvious tendency to describe brain function in terms of the dominant technology of the times.

Not everyone bought into the Galenic model. Most notably, the first great life scientist of the Renaissance (other than Leonardo), Andreas Vesalius, stated in his revolution-sparking masterpiece, the *Fabric of the Human Body* (1543), that the ventricular theory was unsubstantiated and unlikely and that nerves did not look hollow to him. In fact, he refers scornfully to the very drawing reproduced in Figure 5.1. Nevertheless, he was unable to offer any alternative explanations, theories, or models. The start of the third generation of general theories was left to Thomas Willis, who published the first separate volume on the nervous system, the *Cerebri Anatomie*, in 1664. Here, Willis transferred the functions relegated during medieval times to the ventricles back into the brain substance itself. In doing so, he suggested that the cerebral nuclei or basal ganglia (it was he who named them the *corpus striatum*) receive all of the various sensory modalities and thus correspond to the "sensus communis." He also proposed that the corpus callosum generates imagination, that the cerebral cortex is the seat of memory, and that together they control voluntary behavior. In contrast, he suggested that involuntary behavior and the vital functions of the body are controlled by the cerebellum.

As important as Willis's speculations were in shifting attention back to the brain substance itself, they were, after all, only speculations. How the various parts might actually work as a system was left extremely vague because the functional difference between gray and white matter wasn't even known at the time—so memory could be assigned to the cortical gray matter, and imagination to the subcortical white matter. The real breakthrough was provided by two great French experimentalists in the first half of the nineteenth century, François Magendie and Marie-Jean-Pierre Flourens. We have already met and discussed Magendie (Chapter 4). In 1822 he demonstrated experimentally that sensory information enters the spinal cord through the dorsal roots, whereas motor commands to the muscles for behavior leave the spinal cord through the ventral roots (Chapter 4).

In Magendie's nervous system, sensory information enters the spinal cord via one set of nerve fibers and its influence is reflected back out of the spinal cord through another set of nerve fibers to control the muscles (Fig. 5.3). Without any inkling about underlying cellular mechanisms, Magendie demonstrated that there are sepa-

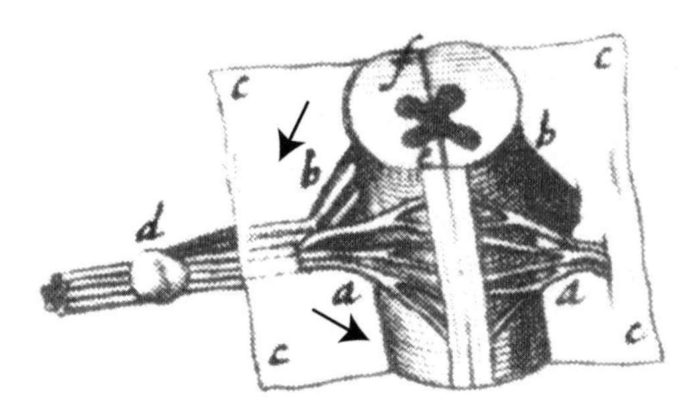

FIGURE 5.3 *This illustration of Magendie's circle is from the first monograph on the spinal cord, by Gerard Blasius in 1666. In the* Anatome Medullae Spinalis et Nervorum inde Provenientium *Blasius reported the discovery of the dorsal and ventral spinal roots, along with the* H*-shape of the spinal cord gray matter. Arrows added to indicate direction of information flow. Key: a, dorsal root; b, ventral root; c, dura; d, spinal ganglion, e, dorsal median sulcus; f, ventral median fissure.*

rate sensory and motor systems and that they have an obligatory interaction within the central nervous system. In the 1830s the pioneering British neurophysiologist Marshall Hall named this arrangement the *reflex arc*. It is a fundamental part of all subsequent models of basic nervous system organization.

Flourens did for the brain what his teacher Magendie did for the spinal cord, and their results were presented almost simultaneously. Based on the first systematic experimental analysis of brain function (using experimental lesions), Flourens concluded that the cerebral hemispheres are the seats of sensation and intelligence, the cerebellum of motor function, and the hindbrain of vital functions. Flourens's stature was so great that by 1840 he was able to defeat Victor Hugo for a lone chair in the French Academy. As we shall see later, this authority was not necessarily good—based on his experimental results, Flourens strongly opposed the idea of cortical localization.

A recurring theme in the history we have been considering so far is functional localization in the nervous system. In fact, it would be hard to think of a better organizing principle for the history of neuroscience than the more and more accurate localization of different functions to distinct parts of the brain. However, the work of the experimentalists acquired a whole new interpretation when the full implications of the cell theory were finally applied to the organization of neural systems by a small though brilliant group of neuroscientists—with Cajal at the helm—toward the end of the nineteenth century. This was the neuron doctrine and its corollary, functional polarity, and the way they have been applied to simpler nervous systems was a theme of Chapter 2. It is now time to see how they apply to the basic organization of the vertebrate nervous system, and more specifically to the mammalian nervous system (including humans).

REFLEX AND VOLUNTARY CONTROL OF BEHAVIOR

The first circuit diagrams of the nervous system based on the cell theory (the neuron doctrine) were published by Cajal in 1890, (Fig. 5.4), and in an important way they explained the results of both Magendie and Flourens. At the level of the spinal cord, the axon of a dorsal root ganglion cell transmits sensory information into the spinal cord via the dorsal roots. This information goes directly, or is relayed by another neuron (an interneuron), to a motoneuron, whose axon leaves the spinal cord through a ventral root before innervating a muscle fiber. This describes the cellular architecture of the simplest reflex arc.

However, Cajal made two other fundamental observations. First, he showed that psychomotor neurons in the cerebral cortex also send their axons to motoneurons in the spinal cord. So, motoneurons actually have at least two functionally different sources of axonal inputs or synapses: reflex inputs from sensory neurons and voluntary inputs from cerebral cortical neurons. His other fundamental observation was that, in general, sensory information bifurcates in the central nervous system: part of it goes to the motor system for initiating reflex responses, and part of it goes to the psychomotor or cognitive neurons for influencing voluntary responses. This organization is presented schematically in Figure 5.5.

Cajal's is probably the most compelling and concise model of basic nervous system organization ever presented, and it is worth

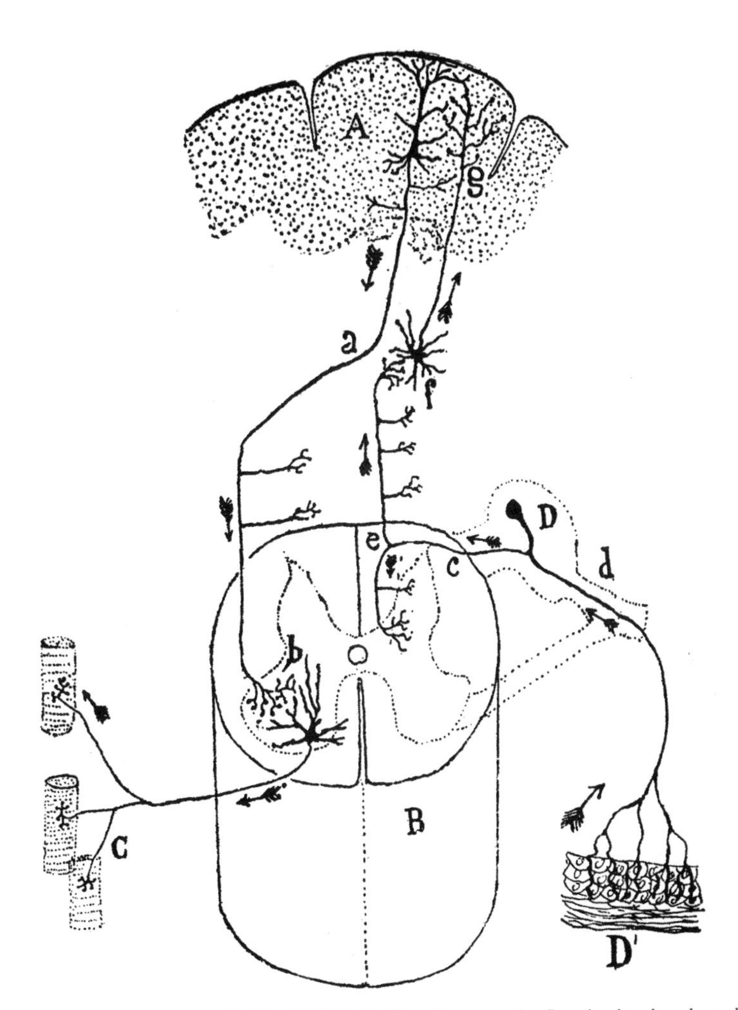

FIGURE 5.4 *In 1890 Cajal published the first diagrams of reflex circuitry based on the neuron doctrine. In this diagram he shows how sensory information from the skin (D′) passes along the sensory fiber (d) of a dorsal root ganglion cell (D), and then its root fiber (c), entering into the spinal cord (B). The sensory root fiber bifurcates at (e): some collaterals of bifurcation branches end in the spinal cord, and the ascending bifurcation branch extends as far as the medulla (f), where it ends on a cell that eventually sends information to the cerebral cortex (g; Cajal was unaware of a relay through the thalamus). A second source of inputs to the spinal cord (b) arises from pyramidal neurons in the cerebral cortex (A). Sensory and cortical inputs to the spinal cord influence motoneurons that send an axon to striated motor fibers (C). Golgi method. From S.R. Cajal,* Les Nouvelles idées sur la structure du système nerveux chez l'homme et chez les vertebrés. *(Reinwald: Paris, 1894). See English translation by N. Swanson, and L.W. Swanson,* New Ideas on the Structure of the Nervous System in Man and Vertebrates *(MIT Press: Cambridge, 1990).*

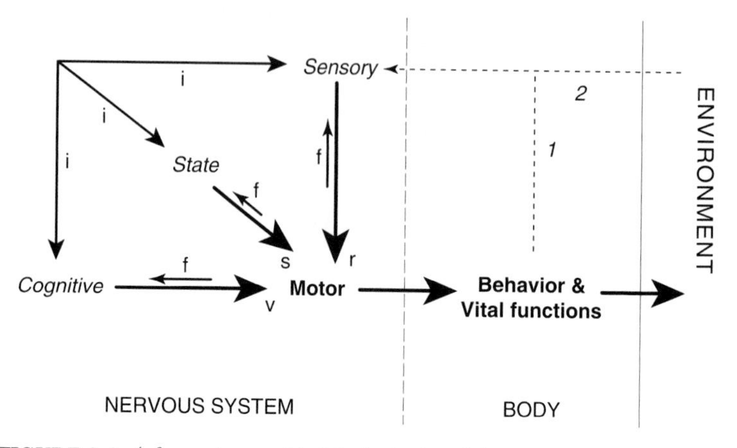

FIGURE 5.5 *A four systems model of the basic plan of the nervous system is shown diagrammatically. One system, the motor system, controls behavior; in other words, behavior is a function of motor system output. In turn, the motor system is controlled by three classes of inputs—from the sensory, behavioral state, and cognitive systems. Direct sensory inputs to the motor system mediate reflex behaviors (r; see Fig. 5.3), inputs from the cognitive system mediate voluntary behavior (v), and inputs from the third system mediate state control influences (s). The motor system influences vital functions within the body (physiological behaviors), as well as the external environment via the skeletomotor system (behavior). These influences on the internal and external environments provide sensory feedback signals to the nervous system (1 and 2, respectively). There is anatomic evidence to suggest that the sensory, behavioral state, and cognitive systems are interconnected with one another (i) and that the motor system as a whole also provides feedback signals to the three other systems (f).*

mulling over a little further before moving on. The first premise, which originally may have been implicit but now should be made explicit, is that the motor system produces behavior, or, put another way, behavior is a function of the motor system. When we examine the behavior of another person or animal, we are observing the effects of the motor system via the motor nerves on the musculoskeletal system. The second premise is that there are two major classes of functional inputs to the motor system: sensory or reflex, and cognitive or voluntary; one originates in the sensory nerves, and the other comes ultimately from the cerebral cortex. Presumably, behavior may be controlled by either or both sources of inputs depending on circumstances. The third premise is that sensory information bifurcates and goes to both the motor system and the cognitive system. Direct sensory inputs to the motor system produce involuntary reflex behaviors when they are strong enough, and inputs from the cerebral cortex mediate the voluntary initiation of the same behaviors. In either case, the same motoneurons control the behavior via their axons to the appropriate muscles.

The perceptive reader may have noticed that the sensory, motor, and psychomotor neurons in Cajal's circuit diagram (see Fig. 5.4) were replaced by terms with much broader meanings in the schematic representation of the diagram (see Fig. 5.5)—sensory system, motor system, and cognitive system, respectively. The justification for doing this is the topic of the next four chapters, but the short answer is that almost all parts of the nervous system can be thought of as parts of systems that control motoneurons, transmit sensory information, or form part of the cerebral hemispheres. But why "almost" all parts?

BEHAVIORAL STATE CONTROL

In light of more recent evidence, there is a third class of inputs to the motor system that needs to be added to the basic wiring diagram proposed by Cajal—a system with intrinsic activity that controls behavioral state (see Fig. 5.5). At the risk of stating the obvious, which we often tend to neglect, the activity patterns of most animals are related in a definite way to the day–night cycle, and in mammals this is governed by fairly regular periods of sleep and wakefulness. This is fundamentally important because the overall pattern of behaviors is very different during sleep and wakefulness. Put another way, the pattern of information flow to the motor system from the sensory and cognitive systems is fundamentally different during wakefulness and during sleep.

There is a neural system that is responsible for switching the overall function of brain circuitry between two radically different states, sleep and wakefulness. And as a corollary for this, it is also responsible for orchestrating less radical differences between various stages of the sleep cycle (for example, deep and rapid eye movement stages), and various levels of arousal while awake. In essence, this system is responsible for controlling behavioral state, and its basic cyclicity is the result of an intrinsically driven clock or clocks, in principle just like there are intrinsic rhythm generators for breathing in the hindbrain and for the heartbeat within the heart itself.

Until recently, it was quite popular to analyze brain function only in terms of stimulus–response relationships. This approach, which was championed especially by a school of psychology known as *behaviorism*, liked to think of the brain as a passive machine, waiting for environmental stimuli to arrive and activate the appropriate response. What this approach chose to ignore was the fact that the brain is a living machine operating all the time. At least three basic findings undermined the behaviorists. First, the brain actually uses just as much, if not more, oxygen when it is asleep as compared to when it is awake and "active." Second, there is a great deal of endogenously generated, intrinsic neural activity. And third, the motor system of embryos is quite active before sensory pathways have even developed to the point of establishing inputs to it! We pointed out in Chapter 2 that most if not all neurons show some level of "spontaneous" activity, which is modulated up or down by synaptic inputs, and it is now clear that there are endogenous rhythm generators as well, some of which control behavioral state.

FEEDBACK

Norbert Wiener formally introduced controllers, feedback, and many other fundamental concepts about systems to biology in his revolutionary book, *Cybernetics.* It was published just after the end of World War II, and it played a major role in establishing the field known today as *computational neuroscience* and one of its practical offshoots, artificial intelligence. We dealt briefly with the idea of behavioral state controllers in the last section, and now it is time to introduce the concept of feedback as it applies to the problem of the nervous system and behavior. Recall the basic plan developed in the last two sections (see Fig. 5.5): behavior (motor system output) is modulated by three classes of input, from the cognitive system, state control system, and sensory system. This provides systems or modules for the voluntary, reflex, and cyclical modulation of behavioral sequences. But what produces activity in these systems, and how are their functions coordinated?

The sensory system is one direct source of input to the motor system, and we can close Magendie's circle of information flow by indicating that the results of behavior feed back into the sensory system (see Fig. 5.5, left-facing arrow, top of figure). The central nervous system is thus constantly informed about what the animal has been doing—a record of its behavior—via feedback from the sensory system. Future behavior is influenced by past experience. We learn from our successes and failures; we try things again if they are positive experiences, or we avoid doing things again if they are negative. Thus, we use feedback from behavior to remember what we have done and to plan what we are going to do. And don't forget that sensory information is also transmitted to the cognitive and state control systems, and thus can alter their output as well.

What all of this implies is that in the awake state there is a constant flow of information from the sensory system to the motor, cognitive, and state control systems and that this flow is modulated by the consequences of behavior. This mode of operation is much different in sleep, when the state controller inhibits the sensory and motor systems, leaving the cognitive system to dream.

This basic plan of nervous system organization is supported by a large body of well-documented anatomical, physiological, and chemical literature, which also shows unequivocally that the functions of all the systems are coordinated by an organized set of connections between them. We have already indicated that sensory information reaches each of the other three systems, and this is also true for the behavioral state system, which sends information (in the form of action potential patterns, as well, perhaps, as "hormonal" signals through the cerebrospinal fluid) to the other three systems. All that remains is to point out similar evidence demonstrating that the cognitive system also projects to each of the other three systems. In short, all four systems—motor, and cognitive, state control, and sensory—share bidirectional connections (Fig. 5.5).

TOPOGRAPHY VERSUS SYSTEMS

The model of the nervous system we are discussing shouldn't be thought of as vertically arranged; it is better to think of it as horizontally arranged. It is not linear (say from rostral to caudal), and it is not hierarchical (say from higher to lower); instead, it is distributed and interactive. Three interacting systems control the motor system and thus behavior; in turn, these systems are controlled interactively by extrinsic stimuli and intrinsic activity. We are dealing with a network, not a hierarchy.

This may seem like an overly simple, if not simplistic, basic plan for the nervous system, but it can be very useful for explanatory purposes if it accounts for known structure and function, and if it accommodates the results of future work. But it bears no obvious resemblance to the basic plan of the nervous system outlined in the previous chapter—a plan using the basic parts that differentiate within the walls of the neural tube. We now come face to face with two seemingly different fundamental plans of the nervous system, one based on *parts or regions*—the cerebral hemispheres, interbrain, midbrain, and so on—and another based on *systems*—motor, cognitive, state control, and sensory. As noted in the preceding chapter, this is a classic dichotomy in anatomy: parts, regions, or topography versus systems. Does one dissect and analyze regions of the body such as the head, trunk, and limbs; alternatively, does one dissect systems: skeletal, muscular, digestive, and so on. The answer is clear: both approaches are valuable: the hand has an obvious function, yet it also has components of many functional systems within it.

However, Figure 5.5 raises another issue: the use of highly schematic diagrams as opposed to the illustration of actual structure or real anatomical relationships. The situation is exactly analogous to comparing the structural organization of the cardiovascular system with Harvey's model of the circulatory system (Fig. 5.6). One is physically accurate, and the other is a simplified, logically correct diagram (which Harvey confirmed experimentally). Again, both approaches are useful and valid.

Nevertheless, when all is said and done, the relationship between the basic systems plan (Fig. 5.5) and the basic topographic plan of the central nervous system (Fig. 4.17)—or even of the nervous sys-

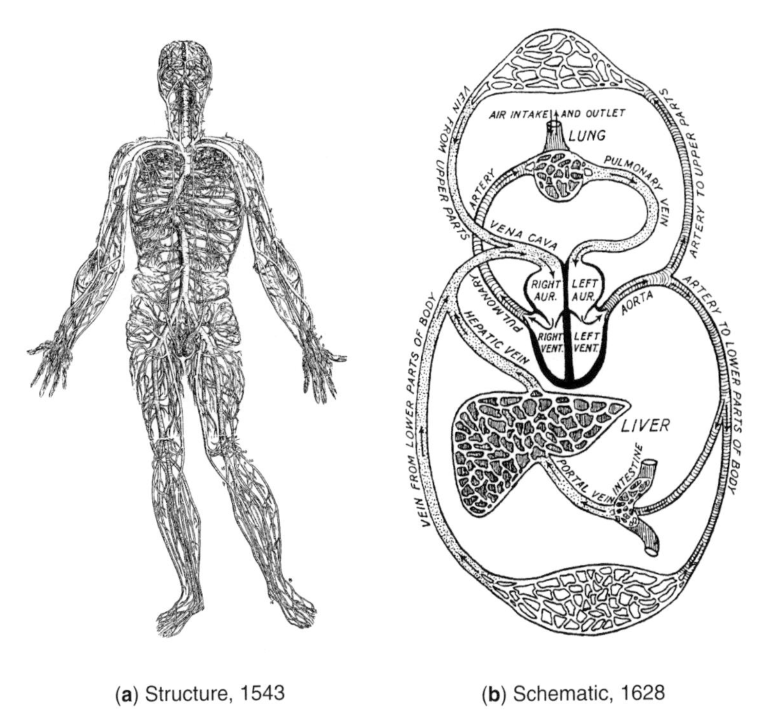

(**a**) Structure, 1543 (**b**) Schematic, 1628

FIGURE 5.6 *Alternative structural (a) and functional (b) ways of illustrating the cardiovascular system. The figure on the left shows the actual structural architecture of the human arterial system and is from Andreas Vesalius's* De humani Corporis fabrica libri septem, *1543. The figure on the right shows the cardiovascular system from a functional point of view, based on the experimental work William Harvey published in 1628. It is from C. Singer,* The Discovery of the Circulation of the Blood *(Bell: London, 1922; repr.: Dawson: London, 1956).*

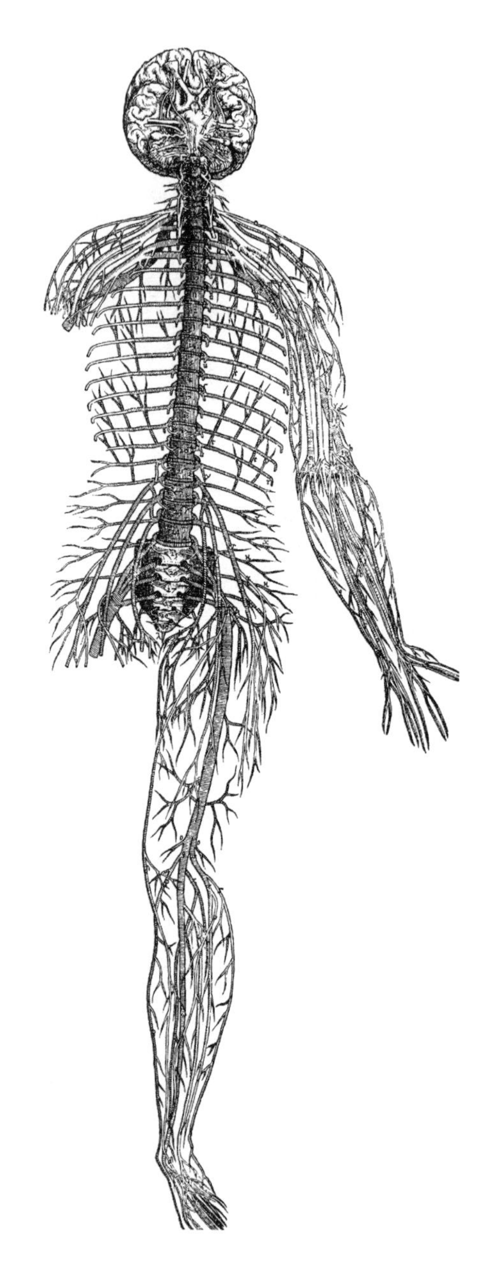

FIGURE 5.7 *This view of the human nervous system as a whole is also from the* Fabrica *of Vesalius (1543). It shows the ventral aspect of the brain, and the spinal cord remains within the vertebral column.*

tem as a whole (Fig. 5.7)—is not straightforward. This problem is actually the major focus of the next four chapters, on each of the major neural systems.

OVERVIEW *Defining Each System*

Two and a half millennia after Aristotle began to formulate general theories about how the body produces behavior through the action of psychic pneuma, we can propose that the nervous system controls behavior via the motor system, which, in turn, is modulated by coordinated inputs from three systems: cognitive for voluntary control, sensory for reflex control, and intrinsic for behavioral state control. And at the cellular level, information flow through this network, which is modulated by feedback, is mediated by an alternating sequence of electrical and chemical events along the axon and at synapses, respectively. But what is the relationship between these four functional systems and the basic structural parts defined as the nervous system develops in the embryo? The answer to this question depends on how the four systems are defined, and thus organized, and this is the topic of the next four chapters. The strategy for defining the systems is time-honored: deal first with the easiest systems to understand because their structural organization and functional dynamics are best understood, and then move to less and less characterized systems, and see what if anything is left unexplained at the end.

READINGS FOR CHAPTER 5

Clarke, E., and Dewhurst, K. *An Illustrated History of Brain Function: Imaging the Brain from Antiquity to the Present*, second edition. Norman: San Francisco, 1996.

Clarke, E., and O'Malley, C.D. *The Human Brain and Spinal Cord: A Historical Study Illustrated by Writings from Antiquity to the Twentieth Century*, second edition. Norman: San Francisco, 1996.

Herrick, C.J. *The Brain of the Tiger Salamander*, Amblystoma tigrinum. University of Chicago Press: Chicago, 1948. Perhaps the most comprehensive and coherent basic plan for the vertebrate nervous system ever published.

Magoun, H.W. Early development of ideas relating the mind with the brain. In: G.E.W. Wolstenholme and C.M. O'Connor (eds.) *The Neurological Basis of Behavior* (CIBA Foundation Symposium). Churchill: London, 1958, pp. 4–27.

Manzoni, T. The cerebral ventricles, the animal spirits and the dawn of brain localization of function. *Arch. Ital. Biol.* 136:103–152, 1998. More than you thought possible to know about the longest lived theory of brain function.

Nauta, W.J.H., and Karten, H.J. A general profile of the vertebrate brain, with sidelights on the ancestry of cerebral cortex. In: F.O. Schmitt (ed.) *The Neurosciences: Second Study Program.* Rockefeller University Press: New York, 1970, pp. 7–26. A modernized, more generalized basic plan à la Herrick.

Sherrington, C.S. *The Integrative Action of the Nervous System.* Scribner: New York, 1906. The masterpiece of neurophysiology and the hierarchical organization of reflexes; reprinted by Yale University Press in 1947.

Swanson, L.W. Cerebral hemisphere regulation of motivated behavior. *Brain Res.* 886:113–164.

Wiener, N. *Cybernetics: Or Control and Communication in the Animal and the Machine.* Herman et Cie: Paris, 1948.

Williams, P.L. (ed.) *Gray's Anatomy*, thirty-eighth (British) edition. Churchill Livingstone: New York, 1995. The authoritative survey of embryonic topology and adult functional anatomy.

6

The Motor System

Coordinating External and Internal Behaviors

The central organs form the connecting medium between all the nerves, or conductors of nervous influence. They act as exciters, or motors of nervous action, in determining the motor nerves to the production of contraction in muscles; and in this their action may be automatic, or voluntary.
—JOHANNES MÜLLER (1843)

We may distinguish two main groups of activities in the vertebrate organism which have determined the general plan of organization of the nervous system: actions in relation to the external world, and internal activities having to do with the processes of nutrition and reproduction.
—J.B. JOHNSTON (1906)

By definition, the motor system is the output of the central nervous system. When I watch you eat a piece of candy, I am actually watching the immediate result of activity in your motor system, which, incidentally, is also controlling motor activity I can't directly see—your swallowing, gut peristalsis, sphincters, and heartbeat. In contrast, the sensory system provides input to the central nervous system, and the behavioral state and cognitive systems are intrinsic to it. We may summarize the basic wiring diagram or plan outlined in the previous chapter by saying that information processing in the central nervous system directs behavior via the motor system, which in turn is controlled by the behavioral state, cognitive, and sensory systems. This chapter is an introduction to the basic organization of the motor system, including its own divisions; the following three chapters go on to discuss in a similar way the three functional systems that control its output. As the quote at the beginning of the chapter from the pioneering American comparative neuroanatomist J.B. Johnston alludes to, there is a long history of dividing bodily functions into two main categories: somatic and visceral. On one hand, there is the "body" or soma with its musculoskeletal system and integument that together deal with the external world; on the other hand, there are all of the various internal, more or less automatic, visceral functions related to digestion, the cardiovascular system, and reproduction.

MOTONEURON CLASSES

There are actually three different motor systems that are quite distinct from one another in terms of both structure and function (Fig. 6.1). A good way to appreciate the differences is to begin by recalling the simple definition of a motoneuron in the nerve net of hydra (Chapter 2). In this case a motoneuron is a neuron that sends its axon to a muscle cell or, more accurately, to a group of muscle cells (see Fig. 2.6). In vertebrates, the best known motoneurons are those that lie in the ventral horn of the spinal cord and send their axon directly to a group of skeletal muscle cells. They are the cells forming the muscles that move joints and certain other structures like the eyes—muscles that are typically attached to bone and under voluntary control but can be activated reflexively as well. They are also known as striated, somatic, or voluntary muscle cells, and while there may be subtle differences in the meaning of these terms, for our purposes they are essentially interchangeable.

The synapses formed by these somatomotor axons on striated muscle cells—synapses known as *neuromuscular junctions*—are the best understood of all synapses because they are so readily amenable to direct physiological manipulation. They are highly differentiated

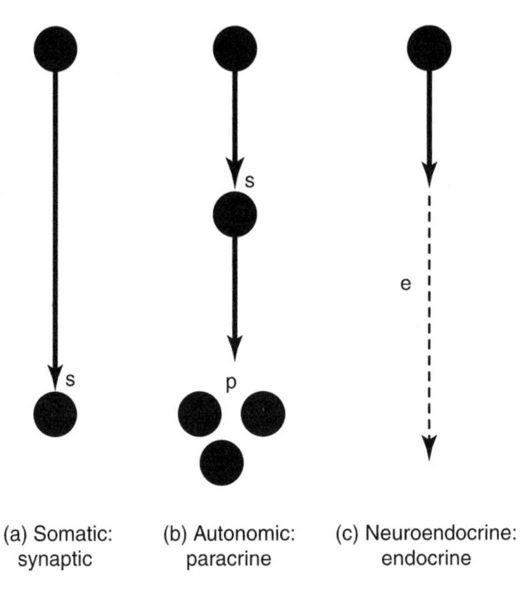

FIGURE 6.1 *The basic arrangement of motoneurons in the three major divisions of the motor system is shown here.* Key: e, endocrine effect; p, paracrine effect; s, synapse on skeletal muscle cell or an autonomic ganglion cell.

both structurally and chemically, with a fairly rigid gap of about 20 to 30 nanometers (0.00002–0.00003 millimeters) between the pre- and postsynaptic membranes—the synaptic cleft that neurotransmitters (acetylcholine and others) diffuse across to influence muscle contractility. The acetylcholine receptors on the postsynaptic muscle membrane are blocked by the South American arrowhead toxin, curare, and many of the classical "nerve gasses," and they are also attacked by autoantibodies in the disease myasthenia gravis. Obviously, paralysis is the result of blocking or inactivating acetylcholine receptors at the neuromuscular junction.

The second motor system, the autonomic system, is concerned with innervating the viscera, and it differs markedly from the somatomotor system in two fundamental ways. First, there is a series of two motoneurons—one in the brainstem–spinal cord and the other in a peripheral autonomic ganglion, which are referred to as the *pre-*

ganglionic and *postganglionic motoneurons* of the autonomic system, respectively. Second, there are two quite distinct divisions of the autonomic system, referred to as *sympathetic* and *parasympathetic*. Broadly speaking, the sympathetic system has short preganglionic and long postganglionic axons, uses noradrenaline as a major postganglionic neurotransmitter, and tends to act as a whole during stressful situations (during the "fight or flight" response to environmental threats). In contrast, the parasympathetic system tends to have long preganglionic and short postganglionic axons, uses acetylcholine as a major postganglionic neurotransmitter, and acts in a localized way that tends to antagonize the sympathetic system (it tends to have a restorative effect and to control digestion).

Most of the visceral organs, along with the blood vessels, receive a dual innervation by the autonomic system, with one division stimulating function and the other inhibiting function, in the process creating a dynamic balance. As a broad generalization, most autonomic system innervation is directed toward three cell types: smooth muscle cells, cardiac muscle cells, and gland cells. Just as skeletomotor neurons were defined for the system that controls striated muscles, so various names have been applied to postganglionic neurons with specific functions. For example, autonomic vasomotor neurons innervate blood vessels and autonomic secretomotor neurons innervate gland cells.

At the level of the autonomic ganglia themselves, acetylcholine is the major neurotransmitter, and as at the neuromuscular junction, nicotinic acetylcholine receptors on the postganglionic membranes mediate the fast synaptic responses. It turns out that most if not all neurotransmitter receptors come in flavors or varieties, and this is true for neuromuscular and autonomic ganglion nicotinic receptors, although in both locations they are blocked by curare and, incidentally, stimulated by nicotine (hence the name). The fact that postganglionic parasympathetic neurons release acetylcholine onto visceral targets, whereas postganglionic sympathetic neurons release norepinephrine, has had vastly important implications for pharmacology—for the development of countless drugs that differentially influence visceral function in one way or another. Just as one example, heart rate is increased by sympathetic stimulation and is decreased by parasympathetic stimulation, or by drugs that interact with acetylcholine and noradrenaline receptors in the heart.

Before moving on to the third motor system, one last feature of autonomic innervation should be noted. Very often, neurotransmitter is not released from a highly specialized synapse like the neuromuscular junction, where the very narrow synaptic cleft assures a one-to-one transfer of information from the presynaptic axon terminal to a specific region of the postsynaptic cell. Instead, autonomic neurotransmitters are often released in the vicinity of groups of cells with appropriate receptors, and the transmitters diffuse to interact wherever cognate receptors are found—hundreds to tens of thousands instead of tens of nanometers away. This arrangement is especially clear for the sympathetic innervation of blood vessels, which works like a garden "sprinkler system" rather than a hose on a tree.

The third motor system, the neuroendocrine system, is centered in the hypothalamus and controls the underlying pituitary gland. Here again there are two major divisions, this time referred to as *magnocellular* and *parvicellular*. The magnocellular division consists of hypothalamic secretomotor neurons that send their axons to the posterior lobe of the pituitary gland, where they release neurotransmitters directly into the blood (the general circulation) to act as classical hormones on a variety of tissues and organs throughout the body. Hormones are molecules that are secreted into, and distributed throughout, the body by the blood to act on any tissues with corresponding receptors. Endocrinology deals with the glands that secrete hormones, as well as the effects of those hormones on target tissues. The parvicellular division of the neuroendocrine system consists of a separate set of hypothalamic secretomotor neurons that sends its axons to the hypothalamic end of the pituitary stalk. Here, in the median eminence, they release neurotransmitter hormones into a system of veins that delivers them to the anterior lobe of the pituitary gland. In turn, these parvicellular neurotransmitter/hormones control the secretion of the all-important anterior pituitary hormones, which are synthesized by five classical cell types that will be discussed later.

Note that neuroendocrine secretomotor neurons exert their classical influence via the blood—in principle, they can influence every cell in the body that expresses an appropriate receptor, assuming of course that the concentration of neurotransmitter/hormone is high enough. The axon terminals of these neurons exert a hormonal influence, which is diametrically opposed to the incredibly focused influence of acetylcholine released by somatomotor neurons at the neuromuscular junction (acting strictly across a 0.02 micron gap on a patch of membrane with an area on the order of a few square microns). Postganglionic axon terminals tend to act over an intermediate range of hundreds of microns, on limited groups of cells. This mechanism has been called "paracrine" in contrast to endocrine or hormonal on one hand and classical synaptic on the other (Fig. 6.1).

INTRODUCTION TO THE SOMATOMOTOR SYSTEM *Flexion*

The somatomotor system mediates the behavior we observe in other people: talking involves controlling the laryngeal muscles, reading involves moving the eyes in a highly patterned way, reaching involves controlling movements of the arm and hand, and so on. The basic mechanics of the system are familiar to everybody: the skeletal system of bones that is moved by the muscles attached to them. To take an example, extend one of your arms straight out, and then flex and extend it a couple of times. Flexion is accomplished by the biceps, "on top" of the arm, whereas extension is accomplished by the triceps, "on the bottom" of the arm. Together, they move the hand and forearm around a hinge called the *elbow joint.* The biceps and triceps are antagonistic muscles that flex or extend the forearm. Let's think a little more about how this arrangement works—how this behavior is mediated.

When your arm is at rest, there is actually tension in all of the muscles—in fact, under normal conditions when you are awake, all muscles are partly contracted; they have tone. This tone is actually controlled by a sensorimotor "proprioceptive" reflex that helps set a "background" level of skeletomotor neuron input to the muscle. The advantage of this situation is that now skeletomotor neuron input to the muscle can be either increased or decreased: muscle tension can be either increased or decreased by inputs from the other three functional systems (behavioral state, cognitive, and sensory).

The biceps is a bundle of many thousands of individual muscle cells that is innervated by a set or group of somatomotor neurons in a specific region of the spinal cord ventral horn. A motoneuron pool is defined as the set of neurons that innervates a particular muscle. Typically they only innervate one muscle (there is little or no divergence via axonal branching to more than one muscle). In contrast, one motoneuron typically branches to innervate more than one muscle cell, called a fiber, within a particular muscle (see Fig. 5.4, C). Thus, as originally defined by Charles Sherrington, a motor unit consists of a particular motoneuron and the set of muscle fibers that it innervates. In adult mammals, only one motoneuron is associated with each muscle fiber. This is basically how the nervous system controls one muscle: by regulating activity in a motoneuron pool dedicated only to that muscle. Left forearm flexion is caused by increased activity in the left biceps motoneuron pool.

Simple physiology experiments show that flexion is more interesting than you might think. When the biceps contracts, the triceps always relaxes at the same time, and vice versa: when the triceps contracts, the biceps relaxes. What this means is that when the biceps motoneuron pool is stimulated to produce flexure, the triceps motoneuron pool (consisting of different motoneurons) is inhibited, which, of course, results in less contraction (the triceps is relaxed). Similarly, when the triceps motoneuron pool is stimulated, the biceps motoneuron pool is inhibited. This arrangement maximizes the efficiency of movement during contraction because the antagonistic muscle is relaxed. The associated neural mechanism is called *reciprocal innervation* of antagonistic muscles.

The discussion thus far presents two basic features of how the motor system controls behavior: individual pools of motoneurons control individual muscles, and all natural movements involve the coordinated activity of more than one motoneuron pool (and so, more than one muscle). Before considering how the coordinated activity of motoneuron pools is achieved, let us take a moment to think about the overall distribution of somatomotor neuron pools.

DISTRIBUTION OF SOMATOMOTOR NEURON POOLS

There is a rock solid foundation on which to build an understanding of the motor system (one of the few in systems neuroscience): the overall distribution of motoneuron pools in the central nervous system. For the somatomotor system in particular, all of the pools are found in the spinal cord, hindbrain, and midbrain (Fig. 6.2).

A continuous longitudinal column of somatomotor neurons extends throughout the length of the spinal cord. These large, multipolar neurons are located in the ventral horn region of the spinal

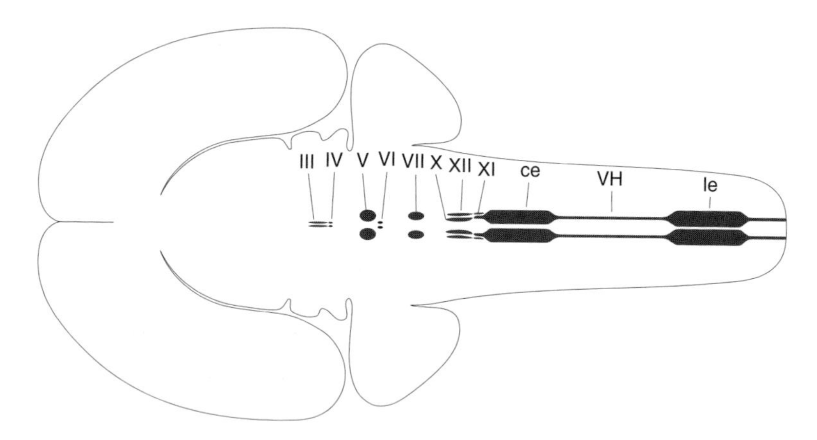

FIGURE 6.2 *The distribution of somatomotor neuron pools is illustrated here on a flatmap of the rat central nervous system (see Fig. 4.17).* Key: *ce, cervical enlargement; III, oculomotor nucleus; IV, trochlear nucleus; le, lumbar enlargement; V, motor nucleus of the trigeminal nerve; VH, ventral horn; VI, abducens nucleus; VII, facial nucleus; X, nucleus ambiguus (of vagus nerve); XI, nucleus of the spinal accessory nerve; XII, hypoglossal nucleus.*

cord gray matter (Chapter 4), and motoneuron pools are arranged in such a way that those for flexor muscles tend to be dorsal to those for corresponding extensor muscles. In levels of the spinal cord related to the limbs (the cervical enlargement for the upper limbs, and the lumbar enlargement for the lower limbs), there is further organization. Motoneuron pools for muscles in the hand are lateral, and pools for muscles progressively closer to the trunk tend to be progressively more medial (Fig. 6.3).

It is curious that whereas there is a continuous distribution of ventral horn motoneurons down the length of the spinal cord, the axons of these motoneurons collect into distinct ventral roots (Fig. 5.3) before joining distinct spinal nerves (Fig. 6.4 and Chapter 9). The number of these spinal nerves varies in different species—for example, in humans there are 31 pairs and in rats 34. One has to wonder whether this regular arrangement of nerves reflects an underlying segmentation of the spinal cord, equivalent perhaps to the neuromeres that are so obvious in the brain during early embryogenesis (Chapter 4). Recent evidence has shown that this is not the case. Instead, it is very clear that the bundling of motoneuron axons into

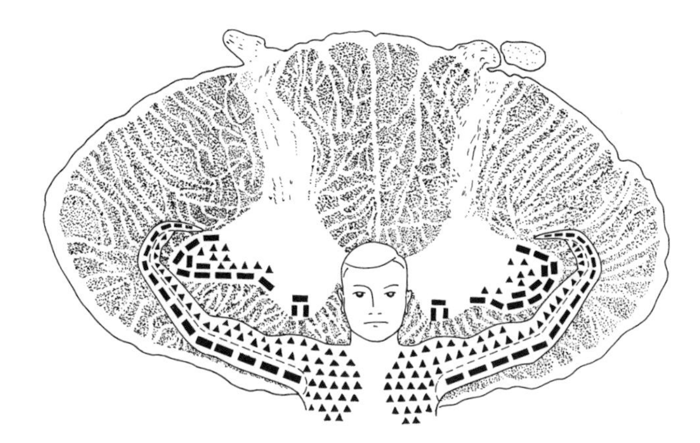

FIGURE 6.3 *The topographic distribution of somatic motoneuron pools as seen in a transverse section through the human spinal cord is illustrated in this drawing. Reproduced with permission from E.C. Crosby, T. Humphrey, and E.W. Lauer,* Correlative Anatomy of the Nervous System *(Macmillan: New York, 1962, p. 73).*

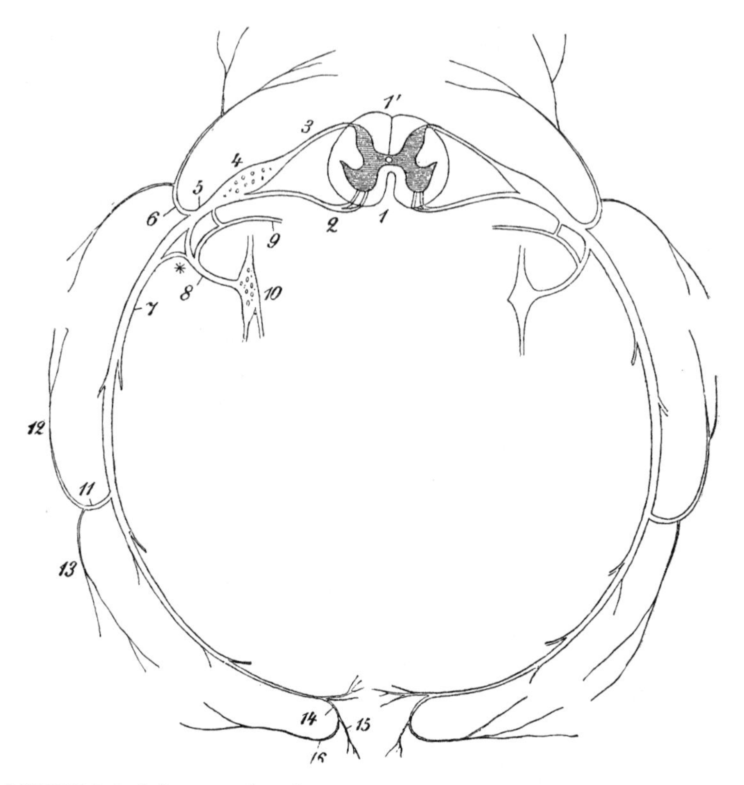

FIGURE 6.4 *A diagram to show the arrangement of the human nervous system in one segment of the thoracic region. At this level the central nervous system is represented in cross-section by the spinal cord, and the peripheral nervous system is represented by the spinal nerves, dorsal root ganglia, and sympathetic chain ganglia. Compare with Figures 3.3, 4.13, 5.3, 5.4, 5.7, and 6.12.* Key: 1, *ventral median fissure;* 1′, *dorsal median sulcus;* 2, *ventral (motor) root;* 3, *dorsal (sensory) root;* 4, *dorsal root ganglion;* 5, *spinal nerve (common trunk);* 6, *dorsal ramus, spinal nerve;* 7, *ventral ramus;* 8, *communicating ramus;* 9, *meningeal ramus; 10, sympathetic chain (paravertebral) ganglion;* 11, *lateral cutaneous ramus;* 12, *dorsal limb,* 13, *ventral limb of* 11; 14, *ventral cutaneous limb dividing into a medial limb* 15 *and a lateral limb* 16. *From L.F. Barker,* The Nervous System and Its Constituent Neurones *(Appleton: New York, 1901).*

discrete ventral roots is due to primary segmentation of the body wall itself. The axons grow into specific regions of embryonic body segments (specifically, the rostral halves of the somites), which are most familiar and easy to understand by remembering the regular arrangement of vertebrae, ribs, and their associated muscles and nerves in the thoracic (chest) region of the body (for the nerves, see Figs. 5.7 and 6.4).

Somatomotor neuron pools in the brainstem send their axons into cranial nerves rather than spinal nerves, and this is not necessarily just a semantic distinction. The embryology of the head and neck is much more complex than that for the rest of the body, and, in the adult at least, there are anatomically separated motoneuron pools, or clusters of pools, for each cranial nerve that innervates striated muscle (Fig. 6.2). With a little squeezing and fudging, neuroscientists have managed to fit all vertebrates into a 12 cranial nerve scheme, which was first proposed by Samuel Thomas von Sömmerring in his medical school thesis of 1778. Which of them are components of the somatomotor system?

Starting with the most rostral somatomotor neuron pools in the adult and working caudally, there are three pairs of cranial nerves that are concerned exclusively with the six muscles that control the movement of each eye. The oculomotor nerve (cranial nerve III) innervates four of these extraocular muscles and thus has four pools of motoneurons in the cell group (the oculomotor nucleus) that generates the nerve. In contrast, the trochlear (IV) and abducens (VI) nerves are very simple: each innervates a single extraocular muscle, and so the corresponding motoneuron pools in the brainstem (in the trochlear and abducens nuclei, respectively) contain only one motoneuron pool. The oculomotor and trochlear nuclei and nerves are within the midbrain, whereas the abducens nucleus and nerve are within the adjacent, rostral end of the pons. We will turn in the next section to the question of how activity in these three sets of nerves is coordinated to produce coordinated movements of the two eyes. This problem is similar in principle to mechanisms underlying the reciprocal innervation of antagonist muscles across a joint, referred to earlier in the chapter.

The muscles for chewing and for moving the jaw in general are innervated by the prominent motor nucleus of the trigeminal nerve (cranial nerve V), which lies in the pons and sends its axons through the motor root of the trigeminal nerve. The facial nucleus comes next. It lies in the caudal pons or rostral medulla (or both) and gives rise to the facial nerve (VII), which plays an exceptionally important role in nonverbal human communication—especially in the expression of emotional state. However, it is also very important in all mammals because, for example, it controls the muscles of the lips. When they are paralyzed, many animals cannot eat properly because food tends to drop out of the mouth. Think about your predicament after the dentist has anesthetized your "mouth"—including your lips! From an embryological point of view, the motor trigeminal (masticatory) nucleus innervates muscles associated with the first pharyngeal or branchial arch, and the facial nucleus innervates muscles associated with the second branchial arch (refer back to Figs. 4.2 and 4.4).

Next we come to a motor cell group with a curious name, the nucleus ambiguus (the "ambiguous nucleus" because it was difficult to identify in the early days). Structurally, it is unusual inasmuch as part of it contains pools of skeletomotor neurons whose axons travel through two different cranial nerves. The most rostral pool innervates the stylopharyngeus, a tiny muscle that helps elevate the pharynx during swallowing and speech, and its axons travel through the IXth cranial nerve (the glossopharyngeal). However, the rest of the skeletomotor neurons in the nucleus ambiguus send their axons into one of the most complex and important nerves in the body, the vagus nerve (cranial nerve X). The skeletomotor neuron pools whose axons travel through the vagus nerve from the nucleus ambiguus innervate the larynx, and thus mediate speech in humans. In addition, they innervate the constrictor muscles of the pharynx, which are an integral part of the later stages of swallowing that are under reflex control. The motor component of the glossopharyngeal nerve is associated with the third branchial arch, whereas the vagus nerve is associated with the remaining arches.

The last two, most caudal, motor nuclei traditionally associated with cranial nerves are a bit confusing. One of them is the motor nucleus of the spinal accessory nerve (called cranial nerve XI). It has two motoneuron pools that are centered in the ventral horn of the first five or so cervical levels of the spinal cord, and the nerve that they generate innervates two muscles of the neck and shoulder re-

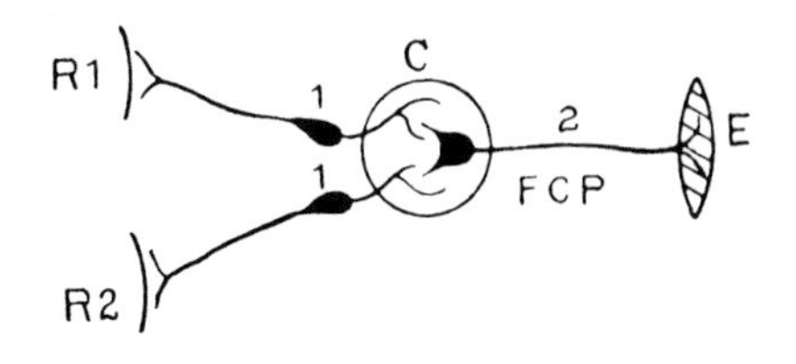

FIGURE 6.5 *The concept of final common pathway (FCP) in the motor system is illustrated in this diagram.* Key: *C, neural center; E, effector; R1 and R2, receptors 1 and 2; 1, inputs; 2, output. From C.J. Herrick,* An Introduction to Neurology *(Saunders: Philadelphia, 1915).*

gion (the trapezius and sternocleidomastoid). In contrast, the hypoglossal nucleus clearly lies in the caudal region of the medulla, and it generates the motor nerve to the complex and fascinating musculature of the tongue (XII, the hypoglossal nerve).

CENTRAL PATTERN GENERATORS
Sets of Motoneuron Pools

The preceding section, dealing with motoneuron pools, was a very straightforward description for the simple reason that the basic structural and functional organization of these neuron groups, which generate the craniospinal motor nerves to the somatic musculature, is well established. Sherrington called them the "final common pathway" of the motor system because they integrate multiple inputs and are the direct output to behavior (Fig. 6.5). In contrast, most of what we can say about the neuroanatomy of the hierarchically organized network that directly controls the patterned output of the final common pathway—the rest of the motor system—is fairly vague and hypothetical. And at this point the whole topic is associated with what can only be described as a chaotic terminology. Fortunately, there doesn't seem to be very much controversy about its general functional organization, leaving aside nomenclature.

What we do know is that the motor system generates constantly changing patterns of activity that require more or less coordination between the hundreds of muscles on each side of the body. This has led to the concept of "central pattern generators" and a variety of related concepts such as central rhythm generators. As we shall now see, experimental evidence indicates that the motor system can be viewed essentially as a hierarchical network of central pattern generators, initiators, and controllers—with the final common pathway at the bottom of the hierarchy, just below the central pattern generators (Figs. 6.6 and 6.7). Thus, the lowest level of the hierarchy is occupied by the motoneuron pools. Think of them as a piano keyboard: a motoneuron pool plays a note (more or less loudly), a specific set of motoneuron pools plays a chord (more or less accurately), and so on until in the end we have a symphony of behavior.

To start let's turn to the real life situation we discussed in the preceding section, flexion and extension of the forearm across the elbow joint. Recall that when the biceps motoneuron pool (in the ventral horn of the spinal cord cervical enlargement) is excited, the arm flexes, and at the same time the triceps motoneuron pool (ventral to the biceps pool) is inhibited through a mechanism of reciprocal inhibition of the antagonist (triceps) muscle. Put another way, an excitatory neural input to the flexor motoneuron pool actually generates a patterned motor response that involves both the flexor and

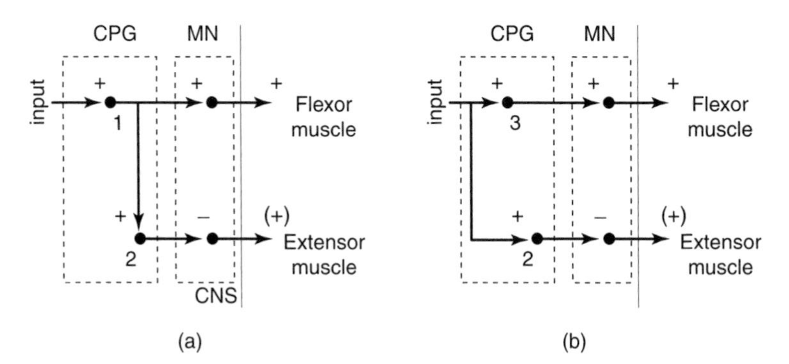

FIGURE 6.6 *Two hypothetical models for circuitry underlying the reciprocal innervation of flexor and extensor muscles are shown in these diagrams. The model in (a) involves two interconnected interneurons (1,2) that form a simple central pattern generator (CPG). The model in (b) also involves two interneurons (2, 3) but in this case the input directly innervates both of them. In (a) and (b) one interneuron is excitatory, +, and the other is inhibitory, −.* Key: *CNS, central nervous system; MN, motoneuron; (+), less excitation.*

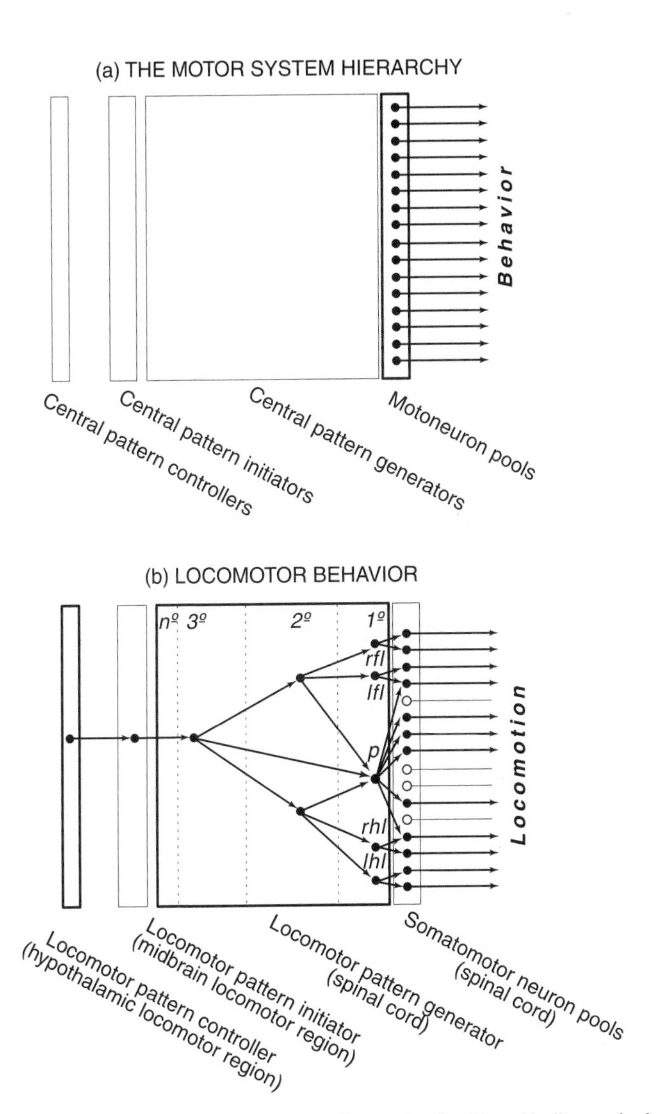

FIGURE 6.7 *The core of the motor system can be thought of as hierarchically organized. Part (a) shows the motoneuron pools at the bottom controlling behavior. Part (b) shows the basic organization of the neural system controlling a specific somatomotor behavior, locomotion. Note the hierarchy of central pattern generators within the spinal cord locomotor pattern generator itself: lfl, left forelimb motoneuron pools; lhl, left hindlimb motoneuron pools; p, postural motoneuron pools; rfl, right forelimb motoneuron pools; rhl, right hindlimb motoneuron pools.*

extensor muscles. The neural mechanism that generates this pattern is, by definition, a central pattern generator.

One simple model of the central pattern generator controlling reciprocal inhibition of antagonist muscles that is consistent with experimental evidence is shown in Figure 6.6a. And we say "model" because it is not yet possible to examine histological sections of the spinal cord under the microscope and point with certainty to the precise neurons that form the simplest of central pattern generators. This is the frighteningly primitive level of understanding we are faced with in explaining the function of vertebrate neural systems. We do know that, to a first order of approximation, all mammalian somatomotor neurons are similar insofar as they excite muscle contraction by releasing the neurotransmitter acetylcholine. When a muscle relaxes, it is because there is less excitation, not because there is active inhibition via an inhibitory neurotransmitter. Thus, when the biceps contracts, the triceps relaxes because its motoneuron pool is inhibited by an inhibitory interneuron (2 in Fig. 6.6a). This inhibitory interneuron is one part of the central pattern generator network for reciprocal inhibition. The other part is an excitatory interneuron (1 in Fig. 6.6a) that receives inputs (excitatory) and then relays them to the flexor motoneuron and to the inhibitory interneuron.

Figure 6.6a is important and should be thoroughly understood because it is a simple example of how neural mechanisms are analyzed and described in terms of networks of excitatory and inhibitory connections. Information flows in one direction along the axons (away from the cell body and dendrites, which are indicated together by a filled circle for the sake of simplicity—the principle of functional polarity discussed in Chapter 2). Another assumption in this model is that the same neurotransmitter (or mixture of neurotransmitters, really) is released from all the axon collaterals and synapses arising from a particular neuron. This is referred to as *Dale's principle*, which was named after Henry Dale, perhaps the greatest physiologist–pharmacologist of the twentieth century, who shared the Nobel Prize with Otto Loewi in 1936 for their role in establishing the chemical nature of synaptic transmission. It is illustrated by interneuron 1, which excites both the flexor motoneuron and inhibitory interneuron 2 (thus inhibiting the extensor motoneuron and relaxing the extensor muscle). Finally, one could look at this as a model of connections between four neurons: two in the central pattern generator and two different motoneurons. However, it is really a model of four *populations* of neurons: the flexor and extensor motoneuron pools and the corresponding groups of excitatory (1) and inhibitory (2) interneurons in the central pattern generator network that innervates the motoneuron pools. Thus, Figure 6.6a is really an abstraction that shows the *minimal circuit* that explains the experimental data and that can perform the task under consideration. The fact that the input in Figure 6.6a ends on an excitatory interneuron and not directly on the flexor motoneuron (pool) is based on experimental results, not on any a priori assumptions.

An alternative model of how the central pattern generator network may be organized is shown in Figure 6.6b. Here, there is also an excitatory interneuron (3) that innervates the flexor motoneuron (pool) and an inhibitory interneuron (2) that innervates the extensor motoneuron (pool). The difference is that in this model the inhibitory interneuron is innervated by a branch of the input fiber to the excitatory interneuron—instead of by a branch of the excitatory interneuron. Whether one or the other (or neither or both) model applies to a particular pair of antagonistic muscles needs to be determined experimentally.

Now we at least have models for the flexor central pattern generator involving the elbow—an excitatory and an inhibitory interneuron connected in stereotyped ways to two functional types of motoneuron (flexor and extensor). What is the next step in the analysis? We need to build a hierarchy of central pattern generators. Instead of voluntarily flexing your elbow, imagine holding out your arm with your eyes closed. If someone were to prick your hand with a pin, there would be an immediate withdrawal of the hand, and, in fact, the whole arm would contract if the prick were strong enough. In neurologic terms, each of the joints in the arm (and the shoulder) would flex in a certain order—a stereotyped sequence of flexor reflexes that are protective in nature: withdrawal reflexes.

As a matter of fact, there is a central pattern generator for regulating antagonistic muscles at each joint in the arm, but there is also a central pattern generator that regulates the sequence of flexion along the joints in the arm. In other words, we have a series of primary or first-order central pattern generators that, in turn, is regulated by a second-order central pattern generator. So the motor system hierarchy we have built so far includes the motoneuron pool layer at the bottom, then a layer of primary central pattern generators that innervate specific subsets of motoneuron pools (for example, related to the flexor and extensor muscles for a particular joint), and then a secondary central pattern generator that regulates the sequence of outputs from the primary central pattern generators. Of course, the actual hierarchy for arm control is much more complex than this: just consider what is involved in controlling the hand alone, with five fingers, each with two or three joints themselves! But the principle remains the same; a hierarchy of central pattern generators coordinates the activity of the many muscles involved in any particular behavior.

We have just been discussing the possibility of a hierarchically organized network of central pattern generators involved in coordinating the movement of an arm—which has a whole series of joints from the fingers, through the wrist and then the elbow, to the shoulder. Now stop and think of an even more complex behavior—*walking*—which involves coordinating movements in all four limbs, along with movements across all the joints within each limb. This is obvious in four-legged animals but is also true in humans, where the arms swing alternately in a very stereotyped way (unless one is crawling). Now let us return to the experimental and clinical data. We know that humans and animals with a completely severed spinal cord (where the spinal cord is completely disconnected from the brain) can still display coordinated locomotor behavior when the limbs are placed on a moving treadmill.

This remarkable fact indicates that there is a locomotor pattern generator situated entirely within the spinal cord, which can be activated by sensory information reaching it from nerves in the feet (and hands). We do not yet know the actual wiring diagram of the locomotor pattern generator network, but we do know that it exists; that it coordinates the incredibly complex sequence of muscle contractions involving rhythmical movement of all four limbs; and that it is an innate, genetically programmed, "hard-wired" circuit. In a very general sense, it must be a hierarchically organized network of more localized pattern generators that control single joints, the series of joints along a particular limb, and finally the rhythmical sequence of limb activations characteristic of locomotion—including different stages of locomotion such as walking and running (Fig. 6.7).

In essence, the locomotor pattern generator is a network of intraspinal interneurons that produces a complex behavior (pattern of muscle contractions) when activated by a combination of behavioral state, cognitive, sensory, or even higher-order motor inputs. Structurally, the pattern generator is located near the set of motoneuron pools that it innervates, so, for example, the locomotor pattern generator is entirely within the spinal cord. Before moving on to the highest levels of the motor system hierarchy (central pattern initiators and controllers), it is worth pausing for a moment to take an inventory of the major behavioral pattern generators at the top of the central pattern generator hierarchy (Fig. 6.8). Taking a broad view, they seem to fall into three rough groups.

One group is obviously concerned with exploratory or foraging behavior. It includes the locomotor pattern generator in the spinal cord; pattern generators for orienting movements of the eyes, head, and neck; and of course supporting both of these, pattern generators for maintaining posture under the constant pull of gravity. Another group of behavior pattern generators appears to be more concerned with behavior after a goal has been approached. Reaching, grasping, manipulating, licking, chewing, and swallowing would be examples here. Finally, other pattern generators have a constant rhythmic activity as long as an animal is alive. A good example of this is the respiratory pattern generator in the ventral medulla and upper

Major behavior pattern generators	Central nervous system location
Breathing	ventral medulla/upper cervical cord
Orofaciopharyngeal movements Facial expression Vocalization Licking, chewing, and swallowing	parvicellular reticular nucleus (dorsolateral hindbrain)
Reaching, grasping, and manipulating	cervical enlargement (spinal cord)
Orienting movements	
Eyes (oculomotor)	dorsal midbrain reticular core
Head and neck	cervical spinal cord
Posture	spinal cord
Locomotion	spinal cord

FIGURE 6.8 *This is a list of the major behavior pattern generators and their approximate locations within the central nervous system.*

cervical spinal cord. Breathing is absolutely dependent on the intrinsic activity of this central pattern generator.

Central pattern generator networks that control the output of a particular set of motoneuron pools to generate a particular behavior are the neural substrate for what the ethologists refer to as fixed action patterns—stereotyped behaviors elicited by specific stimuli that can be more or less complex. There are many examples of a specific behavior (a fixed action pattern) that is released by a specific stimulus (called a sign stimulus). But one of the most interesting conclusions of the ethological analysis is that the central pattern generator that produces a fixed action pattern is itself activated by an innate releasing mechanism—a mechanism in the brain that detects the appropriate stimulus and then releases the appropriate fixed action pattern. In other words, there is a central pattern recognizer that discharges when a specific pattern of stimuli is presented to an animal, and this discharge leads to the activation of a central pattern generator that, in turn, produces a fixed action pattern (behavioral response).

We now continue our analysis of the somatic motor system hierarchy by discussing the regulation of central pattern generators by innate releasing mechanisms or central pattern initiators.

PATTERN INITIATORS AND CONTROLLERS
Drive and Motivation

We have already mentioned that sensory reflex inputs from the ends of the limbs can activate the locomotor pattern generator. In addition, this generator can be activated by experimentally stimulating the midbrain locomotor region, which lies deep to the inferior colliculus (the caudal tectum). That is, the locomotor pattern generator can be activated without somatic sensory reflex inputs from the spinal cord. The midbrain locomotor region is thus a central pattern initiator, and, in turn, it is controlled by regions of the forebrain that establish set-points and other endogenous activity levels. For example, there is a caudal hypothalamic locomotor region (often referred to as the *subthalamic locomotor region*) that appears to play a critical role in the spontaneous or intrinsic activation of the spinal locomotor pattern generator. The hypothalamic motor region is thus a central pattern controller, or part of a central pattern controller, at the top of the motor hierarchy for a particular behavior (Fig. 6.7).

We think this a valid model because animals without a forebrain do not show spontaneous locomotor behavior. However, essentially all of the forebrain except the caudal hypothalamus can be removed, and as long as the rest of the brainstem and spinal cord are intact, animals can display spontaneous, internally generated locomotor activity (Fig. 6.9). This indicates that in some poorly understood way, a hypothalamic locomotor controller provides a certain level of "drive" for locomotor behavior. The experimental evidence would also suggest that other parts of the hypothalamus, together with the cerebral hemispheres, mediate the actual direction and planning of that locomotor behavior, based on the selection of particular goals or goal objects—specific motives and motivational states, if you will.

And so we are faced with even more complexity. The highest level of the motor system hierarchy, that of central pattern controllers, is also hierarchically organized. For example, considerable experimental evidence indicates that the medial nuclei of the hypothalamus are critical nodes in controller networks for at least three specific classes of motivated behavior: ingestive (eating and drinking), defensive (fight or flight), and reproductive (sexual and parental).

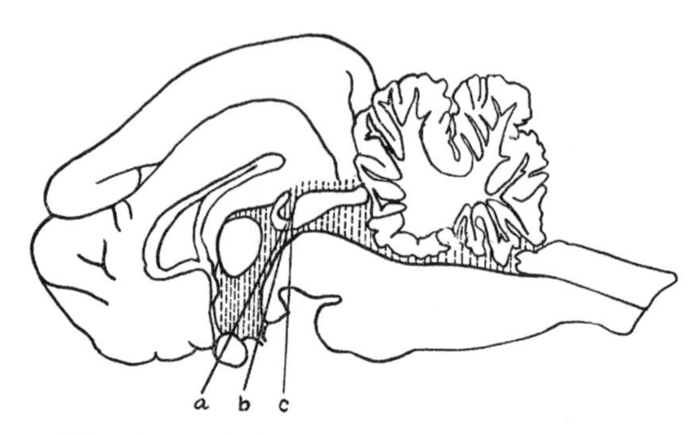

FIGURE 6.9 *When the neuraxis is transected roughly between the midbrain and interbrain (line c), animals display no spontaneous locomotor behavior. They remain immobile until stimulated. In contrast, animals with transection between the interbrain and endbrain (or complete removal of the cerebral hemispheres, and the thalamus) display considerable spontaneous behavior. In fact, they may be hyperactive. This evidence, combined with selective lesions or electrical stimulation of the hypothalamus, suggests that the ventral half of the interbrain (hypothalamus) contains neural mechanisms that regulate set points for locomotor and other classes of motivated behavior. Reproduced with permission of the BMJ Publishing Group from J.C. Hinsey, S.W. Ranson, and R.F. McNattin,* The role of the hypothalamus and mesencephalon in locomotion, *J. Neurol. Psychiat., 1930, vol. 23, p. 17.*

These motivated behavior controllers, in turn, must coordinate essentially all of the simpler behavior pattern initiators listed in Figure 6.8—for example, those associated with foraging behavior and the manipulation of appropriate goal objects.

Finally, we come to the hierarchy of motivated behaviors. This is a bit difficult to establish in mammals, where there is a great deal of flexibility in the sequence of behaviors. In the so-called lower vertebrates and in the invertebrates, however, there is an incredibly sophisticated hierarchy of complex behaviors that are instinctive or genetically programmed, although to a lesser or greater extent they can be modified by experience. The most impressive and persuasive example was probably supplied by the Nobel Prize–winning ethologist, Nikolaas Tinbergen, who described in exquisite detail the sequence of behaviors (the ethogram) associated with the reproductive instinct in a special little fish, the three-spined stickleback (Fig. 6.10). In males,

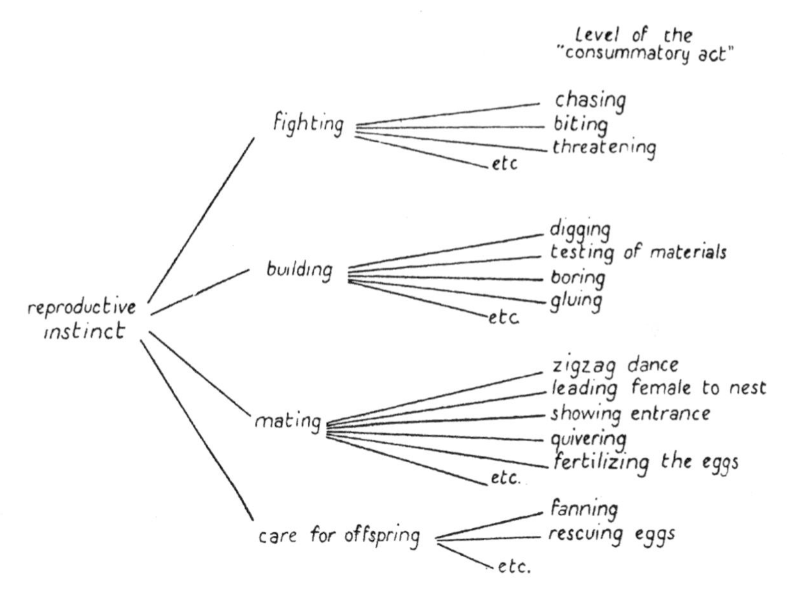

FIGURE 6.10 *This rigidly organized hierarchy of behaviors associated with the reproductive instinct was described for the male three-spined stickleback fish by Nikolaas Tinbergen. If the behavioral sequence is interrupted at any point, none of following behaviors is expressed. Reproduced with permission from N. Tinbergen,* The Study of Instinct. *(Oxford: London, 1951, p. 104).*

the reproductive instinct consists of four sequential behaviors (fighting to establish a territory, building a nest, mating, and care for offspring), and, in turn, each of these behaviors consists of a specific sequence of less complex behaviors.

This example is very instructive in two ways. First, there is a true hierarchical organization because if the sequence of behaviors is interrupted in some way at any point, none of the "downstream" behaviors are expressed. Second, the entire repertoire of behaviors can be turned on or turned off, apparently by influencing the top of the hierarchy. Specifically, the display of the reproductive instinct as a whole is seasonal and is only activated when the length of the day is within a certain range. This ensures that eggs are laid during the appropriate season (spring) for maximal survival and is undoubtedly mediated by the action of gonadal steroid hormones on brain circuits during a certain time of the year (see Chapter 10).

It should be obvious by now that as we move up the motor system hierarchy, away from the motoneuron pools themselves, explanations become more and more vague, and the true situation in terms of neural networks becomes more and more complex. Nevertheless, the basic pattern of motoneuron pools, central pattern generators, central pattern initiators, and then central pattern controllers described for locomotor behavior seems to be well established, and it probably applies to other complex behaviors as well.

Recent evidence suggests that a longitudinal column of distinct cell groups in medial regions of the upper brainstem controls the expression of motivated (or goal-oriented) behaviors and the exploratory (or foraging) behaviors that go along with them (Fig. 6.11). As mentioned, the rostral segment of this column in the hypothalamus has controllers for the three basic classes of goal-oriented behaviors common to all animals, whereas the caudal segment has controllers for the exploratory behavior used to obtain any goal object. Rostrally, at least part of the control mechanism for ingestive behaviors (eating and drinking) is represented in the descending division of the paraventricular nucleus; the control mechanism for reproductive behaviors includes the medial preoptic nucleus, the ventrolateral part of the ventromedial nucleus, and the ventral premammillary nucleus; the control mechanism for defensive (fight or flight) behavior includes the anterior hypothalamic nucleus, the dorsomedial part of the ventromedial nucleus, and the dorsal premammillary nucleus.

The caudal segment of the behavior control column begins with the mammillary body. At least the lateral mammillary nucleus is involved in signaling in which direction the head is pointed (the function of the medial mammillary nucleus is not yet clear). The caudally adjacent reticular part of the substantia nigra is involved in controlling orienting movements of the eyes and head via projections to the superior colliculus. Finally, we come to the ventral tegmental area and caudal end of the midbrain reticular nucleus, both of which appear to be involved in controlling locomotor behavior via mechanisms that

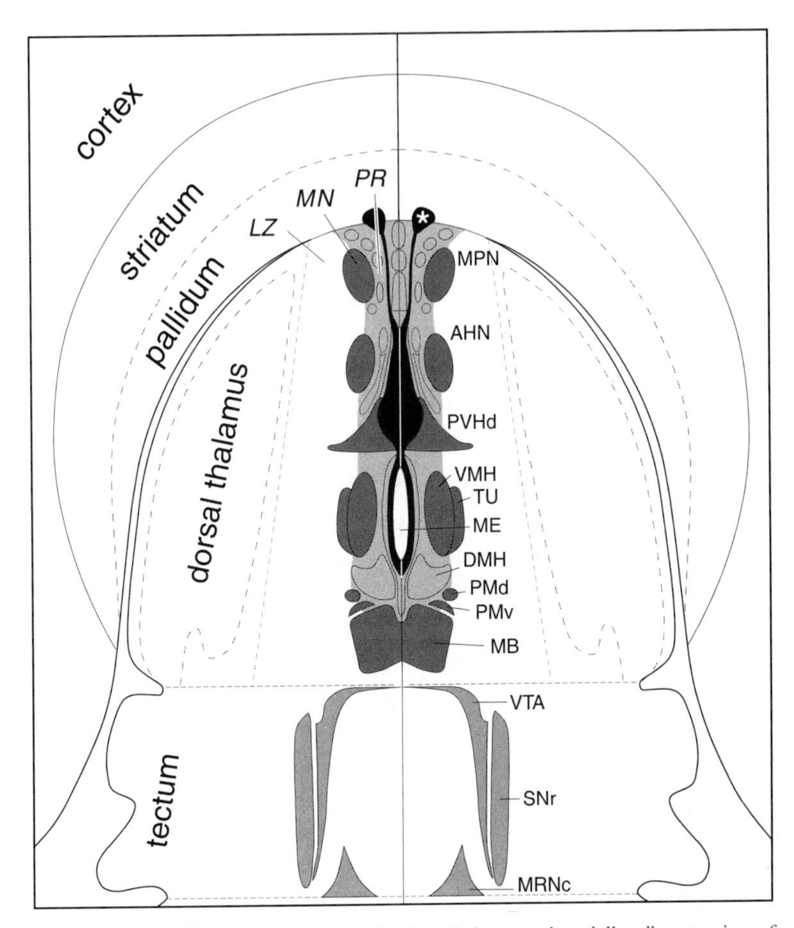

FIGURE 6.11 *The basic organization of the hypothalamus and caudally adjacent regions of the midbrain is shown in this flatmap. The black strip on either side of the midline represents the neuroendocrine motor zone (in the right rostral tip, the white asterisk, *, indicates the GnRH region). The light gray region between the neuroendocrine motor zone and the medial nuclei (MN) is the periventricular region (PR), which contains a visceromotor pattern generator network and the suprachiasmatic nucleus (the master circadian clock in the brain). The medial nuclei and caudally continuous ventral tegmental area (VTA), reticular part of the substantia nigra (SNr), and caudal midbrain reticular nucleus (MRNc) constitute the behavior control column discussed in the text. The hypothalamic lateral zone (LZ) lies lateral to the medial nuclei.* Key: *AHN, anterior hypothalamic nucleus; DMH, dorsomedial hypothalamic nucleus; MB, mammillary body; ME, median eminence; MPN, medial preoptic nucleus; PMd, dorsal premammillary nucleus; PMv, ventral premammillary nucleus; PVHd, descending division of the paraventricular nucleus; VMH, ventromedial hypothalamic nucleus. Adapted with permission of Elsevier Science from L.W. Swanson, Cerebral hemisphere regulation of motivated behavior,* Brain Res., *2000, vol. 886, p. 122.*

remain to be clarified. Along with the nucleus accumbens, the ventral tegmental area appears to regulate the amount of locomotor behavior, perhaps as a component of the subthalamic (or hypothalamic) locomotor region. The caudal midbrain reticular nucleus has been called the midbrain extrapyramidal area by D.B. Rye and colleagues, may correspond at least in part to the midbrain locomotor region.

THE AUTONOMIC MOTOR SYSTEM

Our basic understanding of autonomic motor system organization was laid out through the brilliant work of two English neuroscientists, Walter Gaskell and John Langley, toward the end of the nineteenth century. As mentioned in the discussion of motorneuron classes in this chapter, the autonomic system—which has also been referred to as the *involuntary or visceral system* (in contrast to the voluntary or somatic system just reviewed)—is characterized by two sequential motoneurons: one motoneuron is in the central nervous system and is called preganglionic; the other is in a peripheral autonomic ganglion and is called ganglionic (Fig. 6.12). The general distribution of these motoneuron pools is illustrated in Figure 6.13. Note that preganglionic sympathetic neurons are all found in and near a thin column referred to as the *intermediolateral* (or sometimes, *lateral*) *column* that extends down the thoracic and upper lumbar regions of the spinal cord. The intermediolateral column is separate from, and dorsolateral to, the somatomotor neuron pools in the subjacent ventral horn (Fig. 6.12). In contrast, preganglionic parasympathetic neurons—which typically mediate antagonistic effects to those of the sympathetic system—are found in brainstem nuclei and in sacral levels of the spinal cord (in the intermediolateral column), rostral and caudal to the sympathetic column.

Preganglionic autonomic motoneurons in the brainstem send their axons into several of the cranial nerves, where they course to parasympathetic ganglia in or near the organ that they innervate. Parasympathetic influences on the eye (control of pupil diameter and lens accommodation) are mediated by the oculomotor nerve (III),

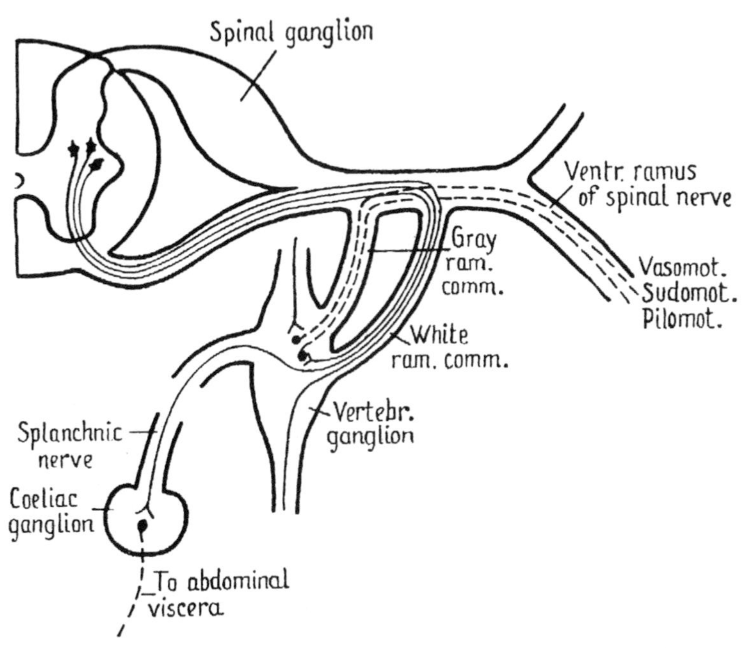

FIGURE 6.12 *This diagram shows the basic arrangement of preganglionic and postganglionic fibers in the sympathetic component of the autonomic motor system (see Fig. 6.1). In the spinal cord, preganglionic fibers arise from neurons in and near the intermediolateral (sometimes just called lateral) column and then course through a ventral root to a paravertebral ganglion (Vertebr. ganglion; see also Fig. 6.4), via a tiny offshoot of the mixed spinal nerve referred to as a white communicating ramus because the axons are myelinated and thus have a white appearance. The postganglionic fibers, which are shown as dashed lines, join the mixed spinal nerve through a separate gray communicating ramus. Most of these fibers are unmyelinated, hence the name. Some preganglionic fibers extend through the paravertebral ganglia to join the splanchnic nerves and end in prevertebral ganglia (such as the celiac ganglion). Reproduced with permission from A. Brodal,* Neurological Anatomy in Relation to Clinical Medicine *(Oxford: London, 1948, p. 347).*

and the preganglionic axons come from a tiny cell group adjacent to the midbrain oculomotor nucleus, the Edinger–Westphal nucleus. Salivation and crying (tear secretion) are mediated by the salivatory nuclei in the medulla, whose axons travel through the facial (VII) and glossopharyngeal (IX) nerves. Finally, two medullary cell

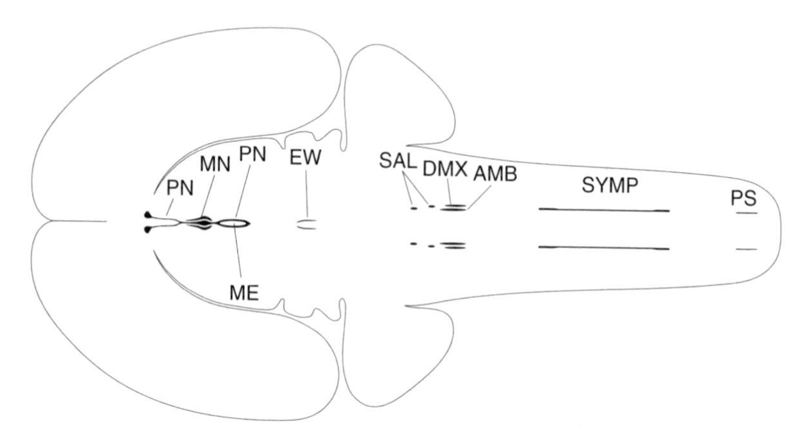

FIGURE 6.13 *The distribution of visceromotor neuron pools displayed on a flatmap of the rat central nervous system. Motoneuron pools associated with the parasympathetic division of the autonomic system include the Edinger–Westphal nucleus (EW), salivatory nuclei (SAL), dorsal motor nucleus of the vagus nerve (DMX), nucleus ambiguus (AMB), and sacral intermediolateral column (PS). The sympathetic division is associated with the intermediolateral column in thoracolumbar levels of the spinal cord (SYMP). Motoneuron pools associated with the neuroendocrine system may be divided into magnocellular (MN) and parvicellular (PN) divisions that are centered in the hypothalamus and associated with the median eminence (ME) (see also Figs. 6.11 and 6.14).*

groups—the motor nucleus of the vagus nerve and part of the nucleus ambiguus—send preganglionic fibers through the vagus nerve (X) to parasympathetic ganglia innervating the heart, stomach, small intestine, and upper colon. Critical parasympathetic inputs to pelvic viscera, including the bladder, lower colon, and genitalia, are associated with preganglionic neurons in sacral levels of the spinal cord.

Sympathetic preganglionic neurons in the thoracolumbar region of the spinal cord send their axons to two classes of peripheral sympathetic ganglia. The first are called the *sympathetic chain ganglia* for a very simple reason: they form a longitudinal chain of ganglia interconnected by bundles of axons. They are a little like two strings of pearls extending along either side of the vertebral column (Fig. 6.12), from the base of the skull to the coccyx. They are reminiscent of the ventral nerve cords in the more advanced invertebrates and are referred to as the *sympathetic trunks*, or *ganglionated cords*. The second class of sympathetic ganglia lie much farther from the spinal cord, in irregular masses associated with visceral branches of the aorta (Fig. 6.12). They supply the abdominal and pelvic viscera, and the preganglionic fibers are carried by splanchnic and pelvic nerves. Most parts of the body are supplied with abundant postganglionic sympathetic nerve fibers (the central nervous system is a notable exception, except that it may have its own version in the guise of a remarkable noradrenergic cell group in the dorsal pons, the locus ceruleus or "blue spot").

It is probably worth commenting on the striking fact that the autonomic motor system has a double output (pre- and postganglionic), whereas the somatomotor system has a single output (the somatomotor neurons). What, if anything, does this have to do with the fact that one system controls the viscera and the other the voluntary muscles (the soma)? The simple answer is that whereas one particular somatomotor neuron sends its axon to one specific muscle, one preganglionic autonomic motoneuron can send axon collaterals to a number of different autonomic ganglia, and one postganglionic axon can branch to innervate a number of different organs or parts of organ systems. This is particularly true for the sympathetic nervous system, which actually derives its name from ancient observations that responses in widely separate viscera throughout the body often can be surprisingly coordinated—they are "in sympathy."

The dual, typically antagonistic, autonomic innervation of the body is highly organized, and there are stereotyped patterns of activity associated with specific behavioral states like exercise, fight or flight, hunger, and sleep. Probably the most famous and dramatic is the emotional excitement and generalized sympathetic discharge aroused in animals that are faced with extreme danger, like the sudden appearance of a predator. In the 1920s the famous Harvard physiologist Walter Cannon studied this "fight or flight response" extensively. He showed that all of the coordinated sympathetic responses that accompany it are directed toward supplying as much energy to muscles as possible, sharpening the sensory modalities, increasing heart rate and blood flow—and decreasing functions that are not vital at the moment, like digestion. At the other end of the spectrum, during sleep the sympathetic system is relatively inactive, and the parasympathetic system comes into play, helping to restore energy supplies. As a general principle, Cannon showed that the opposing actions of the sympathetic and parasympathetic systems play a critical role in maintaining homeostasis, or as Claude Bernard had said in the nineteenth century, a relatively constant internal milieu for the body.

The obvious coordination of responses both within each division of the autonomic motor system and between its two divisions strongly implies that there is a hierarchical organization of autonomic central pattern generators for controlling responses in specific sets of preganglionic motoneuron pools, in a way quite analogous to that described for the somatomotor system (Fig. 6.7). Unfortunately, however, we know very little about the organization, or even identity, of such autonomic pattern generators. One exception is a region of the ventrolateral medulla, close to the nucleus ambiguus and salivatory nuclei, that is involved in coordinating various aspects of cardiovascular homeostasis. Not surprisingly, central pattern generators for breathing are also found in this general vicinity (Fig. 6.8).

THE NEUROENDOCRINE MOTOR SYSTEM

As pointed out in the beginning of this chapter, the neuroendocrine motor system is the final common pathway for controlling the output of the pituitary gland—the master gland of the body's endocrine system. Its motoneurons are centered in the hypothalamus (Figs. 6.11 and 6.13), and they fall into two classes; magnocellular, which are associated with the posterior lobe of the gland, and parvicellular, which are associated with the anterior lobe (Fig. 6.14).

The motoneurons of the magnocellular neuroendocrine system are found in the supraoptic and paraventricular nuclei (and scattered in between them), and they send their axons down into the stalk (infundibulum) of the pituitary gland to end in the posterior lobe, where

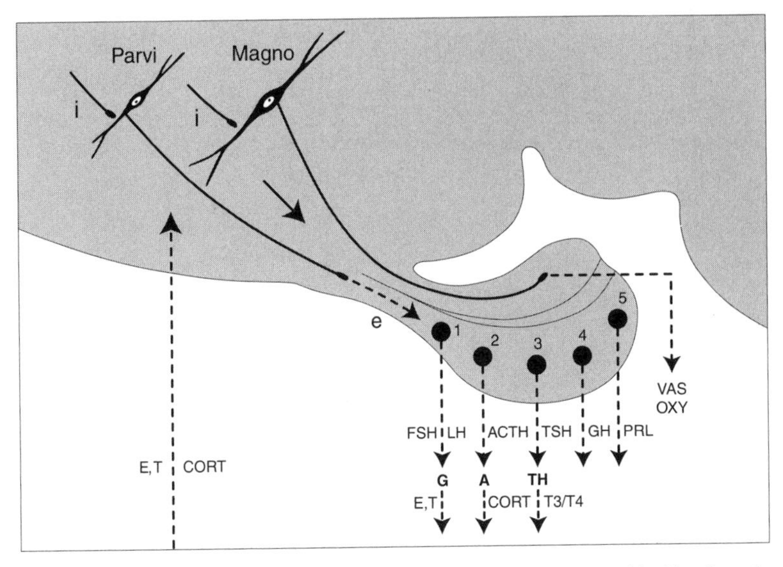

FIGURE 6.14 *The two divisions of the neuroendocrine system are illustrated in this schematic parasagittal view of the hypothalamus and pituitary gland of the rat. Axons of magnocellular neurons (Magno) end in the posterior lobe of the pituitary, where they release vasopressin (VAS) or oxytocin (OXY) in to the general circulation. Axons of parvicellular neurons (Parvi) end in the external lamina of the median eminence (e). Their neurotransmitters are released into hypophysial portal vessels that transport them to the anterior pituitary, where they exert endocrine effects on five classic cell types that, in turn, secrete hormones into the general circulation. Two sources of input to neuroendocrine neurons are shown—neural inputs (i) from other parts of the brain, and endocrine feedback inputs (for example, estrogen, testosterone, and corticosterone).* Key: *A, adrenal cortex; ACTH, adrenocorticotropic hormone; CORT, corticosterone/cortisol; E, estrogen; FSH, follicle-stimulating hormone; G, gonads; GH, growth hormone; LH, luteinizing hormone; PRL, prolactin; T, testosterone; TH, thyroid gland; TSH, thyroid-stimulating hormone; T3/T4, thyroid hormones.*

they release hormones into the general circulation. There are two types of magnocellular neuroendocrine neuron. One of them normally secretes the peptide hormone oxytocin as its neurotransmitter, and this hormone plays a critical part in reproduction: first, it induces powerful contractions of the uterus during parturition; second, it promotes milk release during lactation. The other type normally secretes the closely related peptide hormone vasopressin, which is also called *antidiuretic hormone*, and it plays an important role in controlling blood pressure and water balance. As the names imply, it is a powerful vasoconstrictor (by constricting small arteries, it raises blood pressure), and it is a potent antidiuretic agent (by slowing the formation of urine, it helps retain body water and thus blood pressure). "Magnocellular" means "large-celled," and this is an apt description—they are among the largest neurons in the brain and have an exceptionally high metabolic rate because they are synthesizing an exceptional amount of neurotransmitter. These neurons are also gland cells because they are secreting hormones into the blood, in high enough concentrations to reach and potentially influence all parts of the body. Oxytocin and vasopressin were the first of many neuropeptide neurotransmitters to be purified, characterized, and synthesized, an accomplishment that earned the Nobel Prize for Vincent du Vigneaud in 1955.

The motoneurons of the small-celled, or parvicellular, neuroendocrine system that controls the anterior pituitary are found in and around the ventral wall of the third ventricle (Fig. 6.13). It is hard to overstate the physiological importance of this system because of the hormones secreted by the five classical cell types of the anterior pituitary gland (Fig. 6.14). One of these hormones (ACTH) controls the secretion of glucocorticoids (the steroid hormone cortisol, CORT, in humans) from the adrenal gland cortex. Blood levels of the basic metabolic fuel glucose are regulated by cortisol, which is secreted under all forms of stress. The stress response is critical for survival in the real world. Another hormone, thyroid-stimulating hormone (TSH), regulates the secretion of thyroid hormones, which control metabolic rate throughout the body. A third hormone, growth hormone (GH), is important in establishing body size during maturation, and then regulating metabolism in the adult. A fourth hormone, prolactin (PRL), stimulates milk production after childbirth, and the fifth and sixth are secreted by the final cell type, gonadotropes. The latter are perhaps the most important pituitary cell type of all because they control the secretion of sex steroid hormones (estrogen, E, and testosterone, T) from the gonads—and, in turn, control the female cycle, the sex drive in males and females, and even parental care. Without these functions the species would not survive.

So pituitary hormones control metabolism and body weight, body water and blood pressure, and gonadal and reproductive function. The pituitary is the master gland of the endocrine system, and its own output is controlled by pools of neuroendocrine motoneurons that are centered in the hypothalamus (Figs. 6.1, 6.11, and 6.13). The neurovascular link between the hypothalamus and the anterior pituitary was hypothesized by Geoffrey Harris in the 1940s. However, it took many years before Andrew Schally, Roger Guillemin, and Wylie Vale confirmed it by purifying and synthesizing the peptide neurotransmitter/neurohormones involved in signaling between hypothalamic nerve terminals and anterior pituitary cell types. After 25 years of intense research, they discovered that hormone secretion from a particular cell type in the anterior pituitary is usually controlled by at least one stimulatory hormone and one inhibitory hormone from the hypothalamus. Later histochemical studies with specific antibodies showed that each hypothalamic neurotransmitter/hormone involved in controlling the anterior pituitary is synthesized by a different group of small neurons—the parvicellular neuroendocrine secretomotor neuron pools (Fig. 6.15). Schally and Guillemin were awarded the Nobel Prize in 1977 for their work, after Harris had died and was thus ineligible.

Specific behavioral states are associated with activity in particular sets of somatomotor and autonomic motoneuron pools, and the same applies to the neuroendocrine system. For example, there are different, relatively stereotyped hormonal responses to environments that are too hot or too cold, to strenuous exercise, and to defending one's self from a predator. Thus, one would suspect that there are neuroendocrine central pattern generators, just as there are somatomotor and autonomic motor pattern generators. Actually, such a network has recently been characterized in the medial hypothalamus, in the periventricular region between the neuroendocrine motoneuron pools and the medial nuclei thought to be involved in the highest levels of the somatomotor control sys-

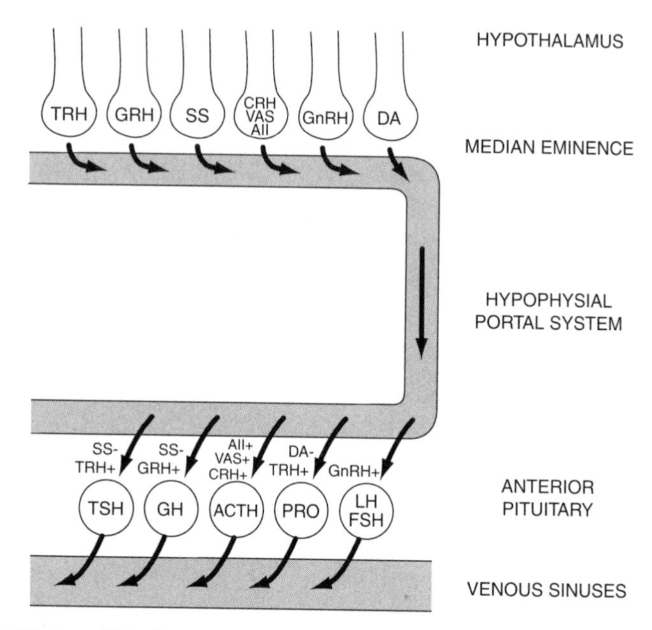

FIGURE 6.15 *This diagram shows how the neurotransmitter hormones are secreted into the hypophysial portal system and transported to the anterior pituitary where they influence the secretion of hormones from the five classical cell types (see Fig. 6.14).* Key: *ACTH, adrenocorticotropic hormone; AII, angiotensin II; CRH, corticotropin-releasing hormone; DA, dopamine; FSH, follicle-stimulating hormone; GH, growth hormone; GnRH, gonadotropin-releasing hormone; GRH, growth hormone–releasing hormone; LH, luteinizing hormone; PRO, prolactin; SS, somatostatin; TRH, thyrotropin-releasing hormone; TSH, thyroid-stimulating hormone; VAS, vasopressin. Adapted with permission from L.W. Swanson, The hypothalamus, in: A. Björklund, T. Hökfelt, and L.W. Swanson (eds.)* Handbook of Chemical Neuroanatomy, *vol. 5 (Elsevier Science: Amsterdam, 1987), p. 19.*

tem (Fig. 6.11, and section on pattern iniators and controllers in this chapter). In addition, the baseline secretion of most pituitary hormones shows an underlying circadian rhythm (over a roughly 24-hour period) and ultradian rhythm (with a cycle time on the order of an hour or two). Neuroendocrine system output as a whole is thus mediated by central pattern generators and central rhythm generators, as is also the case for the somatomotor and autonomic motor systems.

THE CEREBELLUM *Motor Coordination and Learning*

In mammals, the cerebellum ("small brain," as compared with the cerebrum or "large brain"; so-named by Aristotle) is a very conspicuous mass (see the frontispiece and Fig. 4.7) that is attached to the lower brainstem by three pairs of thick fiber bundles, the cerebellar peduncles. As far back as 1664, Thomas Willis guessed that the cerebellum is responsible for controlling what we would refer to as involuntary and visceral motor responses, and almost 350 years later there is still no clear understanding of its function. The only thing that recent textbooks can seem to agree on (after admitting that it is not necessary for either perception or muscle contraction) is that somehow the cerebellum promotes the coordination and fine control of movement by influencing the output of brain motor and cognitive systems—although there are many interesting theories about how it may accomplish these functions.

In view of all this, it is hard not to consider the cerebellum as part of the motor system. But how does it fit into the scheme we have been developing in this chapter? Let us start with the essential structure and circuit diagram of the cerebellum and then go on to consider what its main inputs and outputs are in relation to the rest of the central nervous system. To begin with (Fig. 6.16), the cerebellum has two basic parts: the cortex and the deep nuclei (like the cerebrum). Topologically, the cerebellar cortex is simply a sheet with three layers (granule cell, or deep layer; Purkinje cell, or middle layer; and molecular or superficial layer). In many animals, including all mammals, the area of this sheet has been increased greatly by a process of "corrugation," which leads to innumerable folds (or folia) in the sheet. This cortical sheet forms the surface of the cerebellum, and as the name implies, the deep nuclei lie "underneath" the cortex, embedded in the white matter fiber tracts that carry axons into and out of the cerebellum. This white matter has the quaint name *arbor vitae*, or tree of life; see the human brain in the frontispiece.

The essential circuit of the cerebellum is quite interesting (Fig. 6.16). First, there are two functionally and structurally distinct types

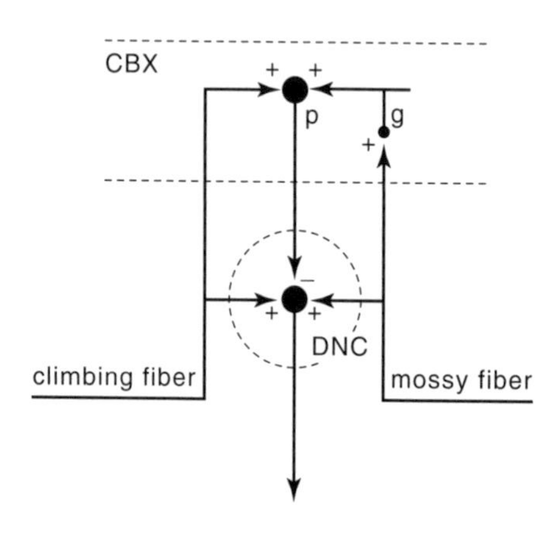

FIGURE 6.16 *This diagram shows the elementary circuit of the cerebellum. For simplicity, interneurons of the cerebellar cortex are not shown.* Key: *CBX, cerebellar cortex; DNC, deep nuclei of the cerebellum; g, granule cell; p, Purkinje cell; +, excitatory; −, inhibitory.*

of specific input to the cerebellum: mossy fibers and climbing fibers, both of which use excitatory neurotransmitters. Second, the output of the cerebellum is generated in the deep nuclei, which are excited by collaterals of both mossy fibers and climbing fibers. Therefore, in a sense, the simplest circuit involving the cerebellum consists of excitatory mossy and climbing fiber inputs to deep nucleus output neurons—which project to the motor system and to the cognitive system via the thalamus (see below). Third, the climbing and mossy fibers go on also to provide excitatory inputs to the cerebellar cortex. And fourth, the cerebellar cortex, in turn, sends an inhibitory projection to the deep nuclei (via Purkinje cells). Thus, the flow of information through the excitatory climbing and mossy fiber collateral inputs to the deep nuclei can be modified by a delayed inhibitory feed forward signal from the cortex.

The elementary cerebellar circuit illustrated in Figure 6.16 provides a beautiful model for considering the importance that timing can have on the function of neural networks. The basic idea here is that deep neurons and Purkinje neurons are in a position to *compare* activity in two classes of input: mossy fiber and climbing fiber. As the simplest possible example, it seems reasonable to expect that impulses arriving simultaneously at a deep neuron from a mossy fiber and a climbing fiber have an additive effect on the deep neuron, whereas unsynchronized inputs should have smaller effects. This type of thinking, combined with a more complete essential model of the cerebellar cortex (that includes inhibitory interneurons), can lead to very fruitful exercises in mathematical modeling and experimental neurophysiological data gathering.

Much more intriguing, however, is the possibility (Chapter 10) that synaptic strength can be increased or decreased by the coincident activation of synapses: in other words, that associative learning can take place. As a matter of fact, Richard Thompson and his colleagues have shown that the circuit illustrated in Figure 6.16 underlies at least some forms of Pavlovian learning, which is also referred to as *classical conditioning*. Recall Pavlov's dogs and how they salivated at the sight of food (an unconditioned stimulus and response). What Pavlov did was to ring a bell just before food was shown to the dog, and the next time the bell was rung alone the dog salivated. Before the pairing, the bell alone did not elicit salivation; after the pairing, it did—it became a conditioned or learned stimulus that produced a conditioned response. The key point was that an ineffective stimulus (the bell) became an effective stimulus after pairing with an unconditioned or already effective stimulus. We now know that the strength of synapses associated with the auditory pathway were strengthened to the point where they were now effective without pairing with the other stimulus. Where does this synaptic strengthening—this learning—take place?

Thompson's group has used a very simple model of Pavlovian learning. The unconditioned stimulus is a puff of air directed toward the eye (cornea), and the unconditioned response is a blink. For the conditioning stimulus a tone is played just before the airpuff. Over a number of trials, there comes to be an eyeblink at the time the tone is played, which actually protects the cornea from the airpuff that is delivered a short time later. In a real sense, a protective reflex has been learned, and it is associated with a previously neutral stimulus (the tone). Now here is the neurobiology (Fig. 6.16): mossy fibers transmit the unconditioned stimulus to the deep nuclei of the cerebellum, and climbing fibers transmit the conditioned stimulus to the deep nuclei. Before training, the climbing fiber input stimulated by a tone is not strong enough to elicit a response in deep nuclei neurons, but after pairing with the unconditioned stimulus input to the deep nuclei, it becomes strengthened to the point where it can elicit a response. The basic memory of a very simple Pavlovian learned response is formed by changing synaptic strength in the deep cerebellar nuclei. From Figure 6.16 it is clear that the same information traveling through the mossy and climbing fibers also reaches the cerebellar cortex somewhat later. It is now known that this extra loop in the circuit helps refine and/or strengthen the basic response that is learned in the deep nuclei.

The essential nature of cerebellar function remains elusive. However, it does appear safe to conclude that "the small brain" is an integral part of the motor system (it is also known to participate in visceromotor responses) and that it plays an important role in motor learning and in fine tuning the coordination between the hundreds of muscles involved in orienting responses, reaching and manipulating, posture, and so on. The cerebellum receives all types of sensory information, either directly from the spinal cord and brainstem or indirectly from the cerebral cortex (via mossy fibers from the pontine gray). After processing in the cerebellum, the resulting information is carried out through the cerebellar peduncles to central pattern generators and central pattern initiators in the brainstem and spinal cord (Fig. 6.16), as well as to the cerebral hemispheres (the cognitive system) via a relay in the thalamus (see Chapter 8).

OVERVIEW *Integration Within and Between Motor Systems*

It is hard to avoid concluding that the core of the motor system is organized in an essentially hierarchical way, with a large set of quite well known motoneuron pools at the bottom level. In contrast, the actual organization of the hierarchy in terms of neuroanatomically characterized networks or circuits remains only vaguely known. One way of summarizing the sketchy functional and structural evidence is illustrated diagrammatically in Figure 6.17. The basic idea is as follows. First, a hierarchy of central pattern generator networks controls the output of the system, the motoneuron pools. A primary central pattern generator innervates a specific set of motoneuron pools

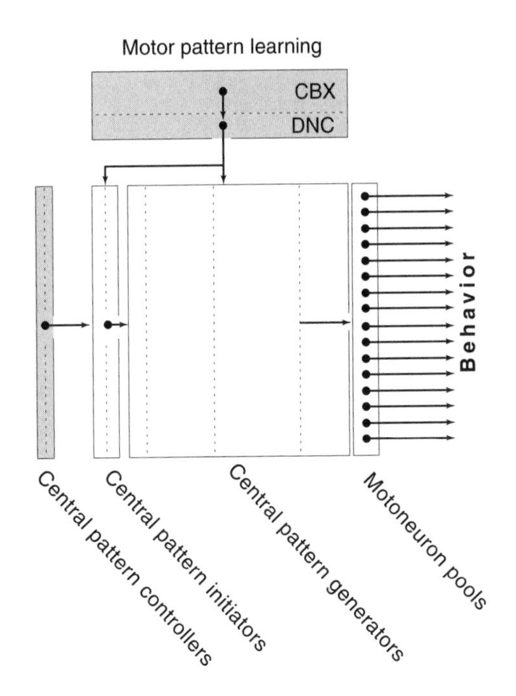

FIGURE 6.17 *This diagram shows the core motor system that is hierarchically organized, with the addition of the cerebellum, which is involved in motor pattern coordination and learning (see also Fig. 6.7).*

and produces a specific pattern of responses (say, contractions in a specific set of muscles) and thus a specific behavior. A secondary central pattern generator network innervates a specific set of primary central pattern generators, producing a specific set of behaviors, and so on. Second, a central pattern initiator projects to the top of a central pattern generator hierarchy for a specific complex behavior (a good example is the spinal locomotor pattern generator, which is activated by a midbrain locomotor pattern initiator). Third, central pattern initiators appear to be under the control of central pattern controllers that are thought to impose set-points or provide intrinsic "drive" or spontaneous activity levels for certain behaviors. Fourth, a great deal of motor coordination and learning appears to occur in the cerebellum, which projects to the central pattern generator and initiator levels of the hierarchy.

In a way, it would appear that two quite different mechanisms control the output of the core motoneuron pool–central pattern generator–central pattern initiator hierarchy. On one hand, there are the central pattern controllers (which for motivated behavior lie in the hypothalamus); on the other hand, there is the motor learning network (a central pattern learner?) in the cerebellum. The cerebellum does not fit easily into a hierarchical model of motor system organization. Taken as a whole, the circuitry outlined in Figure 6.17 for the motor system is organized as a network and not a strict hierarchy because of the cerebellum and its input–output relationships.

As if this were not enough, it is essential that we recall one more wrinkle: there are three motor systems—somatic, autonomic, and neuroendocrine—and their activity is coordinated. In fact, as mentioned before in this chapter, various behavioral states are characterized by more or less stereotyped, coordinated responses in all three systems (Fig. 6.18). Precisely how this coordination is accomplished

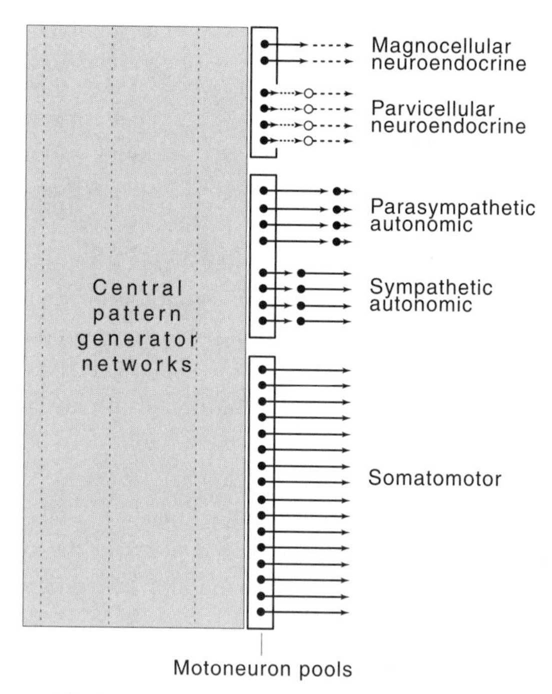

FIGURE 6.18 *This is an overview of the three motor systems—somatic, autonomic (with sympathetic and parasympathetic divisions), and neuroendocrine (with magnocellular and parvicellular divisions). Presumably, central pattern generator networks that remain largely uncharacterized are involved in coordinating appropriate responses in all three systems.*

in neural circuit terms remains to be clarified, but recent anatomical evidence does suggest that the medial hypothalamic central pattern controllers for motivated behavior also project to a medially adjacent visceromotor pattern generator network that coordinates both autonomic and neuroendocrine responses (Fig. 6.11).

In the end, behavior is produced by a core motor system hierarchy that is organized along the lines illustrated in Figure 6.17—with the addition of a cerebellar network that is critically involved in associative motor learning. The output of the motor system as a whole is controlled by three classes of input—from the cognitive, sensory, and behavioral state systems (Fig. 5.5).

READINGS FOR CHAPTER 6

Brodal, A. *The Cranial Nerves: Anatomy and Anatomico-Clinical Correlations*, second edition. Blackwell: Oxford, 1965. This is a model of concise, informative writing.

Brooks, V.B. *The Neural Basis of Motor Control.* Oxford University Press: New York, 1986. Here is a nice overview of general principles.

Eccles, J.C., Ito, M., and Szentágothai, J. *The Cerebellum as a Neuronal Machine.* Springer: New York, 1976. This is a classic book.

Evarts, E.V., Wise, S.P., and Bousfield, D. (eds.) *The Motor System in Neurobiology.* Elsevier: Amsterdam, 1985. This is a nice selection of some 45 short articles on a broad range of topics.

Kandel, E.R., Schwartz, J.H., and Jessell, T.M. *Principles of Neural Science*, fourth edition. McGraw-Hill: New York, 1999.

Kim, J.J., and Thompson, R.F. Cerebellar circuits and synaptic mechanisms involved in classical eyeblink conditioning. *Trends Neurosci.* 20:177–181, 1997.

Kuypers, H.G.J.M. The anatomical and functional organization of the motor system. In: M. Swash and C. Kennard (eds.) *Scientific Basis of Clinical Neurology.* Churchill Livingstone: Edinburgh, 1985, pp. 3–18.

Loeb, G.E., Brown, I.E., and Cheng, E.J. A hierarchical foundation for models of sensorimotor control. *Exp. Brain Res.* 126:1–18, 1999. This is a nice introduction to engineering approaches.

Markakis, E., and Swanson, L.W. Spatiotemporal patterns of secretomotor neuron generation in the parvicellular neuroendocrine system. *Brain Res. Rev.* 24:255–291, 1997.

Nieuwenhuys, R., Voogd, J., and van Huijzen, C. *The Human Central Nervous System: A Synopsis and Atlas*, third edition. Springer-Verlag: New York, 1988.

There are nice, brief summaries of the major functional systems in this beautifully illustrated book.

Orlovsky, G.N., Deliagina, T.G., and Grillner, S. *Neuronal Control of Locomotion: From Mollusc to Man.* Oxford University Press: Oxford, 1999. This is a comparative overview.

Rye, D.B., Saper, C.B., Lee, J.H., and Wainer, B.H. Pedunculopontine tegmental nucleus of the rat: cytoarchitecture, cytochemistry, and some extrapyramidal connections of the mesopontine tegmentum. *J. Comp. Neurol.* 259:483–528, 1987.

Stein, P.S.G., Grillner, S., Selverston, A.I., and Stuart, D.G. (eds.) *Neurons, Networks, and Motor Behavior.* MIT Press: Cambridge, Mass., 1997. Here is an incisive introduction to state of the art thinking about the motor system and its control.

Swanson, L.W. Cerebral hemisphere regulation of motivated behavior. *Brain Res.* 886:113–164.

Tinbergen, N. *The Study of Instinct.* Oxford University Press: London, 1951. This is one of my favorite books; a revelation.

Voogt, J., Jaarsma, D., and Marani E. The cerebellum, chemoarchitecture and anatomy. In: L.W. Swanson, A. Björklund, and T. Hökfelt (eds.) *Handbook of Chemical Anatomy: Vol. 12, Integrated Systems of the CNS, Part III: Cerebellum, Basal Ganglia, Olfactory System.* Elsevier: Amsterdam, 1996, pp. 1–369. This is an exhaustive review.

Williams, P.L. (ed.) *Gray's Anatomy*, thirty-eighth (British) edition. Churchill Livingstone: New York, 1995. It contains an excellent summary of neural systems.

Zigmond, M.J., Bloom, F.E., Landis, S.C., Roberts, J.L., and Squire, L.R. (eds.) *Fundamental Neuroscience.* Academic Press: San Diego, 1999.

7

The Behavioral State System

Intrinsic Control of Sleep and Wakefulness

> The periodical recurrence of sleep and the waking state is, therefore, essentially connected with something in the nature of animals, and is not dependent on the simple alternation of day and night. But the periods of sleeping and waking, in accordance with a pre-established harmony of nature, have been made to agree with those of the earth's revolutions.
>
> —JOHANNES MÜLLER (1843)

If you are like most people, you spend about a third of your time asleep, and while you may not have stopped to think about it, sleep and wakefulness are obviously two entirely different behavioral and mental states. When you are asleep, gentle sensory information doesn't seem to "get in" and your muscles are relaxed—essentially there is no overt behavior, except for breathing and rolling over now and then. No one has come up with a convincing explanation for why we sleep, but there must be one because it has such a long evolutionary history: alternating periods of sleep-like behavior and wakefulness are found in all vertebrates and have even been detected in many invertebrates such as mollusks and insects.

When all is said and done, the sleep–wake cycle is the primary organizer of behavior. As long as you are asleep, there is no overt behavior that one would call "voluntary" in everyday language.

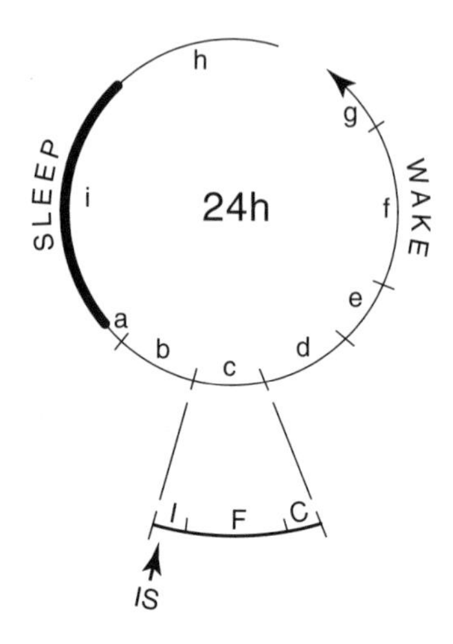

FIGURE 7.1 *The sequence of behaviors (a through i) executed over the course of a typical day is shown in this diagram. Time is shown by the circle with an arrowhead. A complete behavioral episode carried out while awake (c) can be divided into three sequential phases: initiation (I), foraging or procurement (F), and consummatory (C), which includes a satiety mechanism for terminating the episode. IS, initiation stimulus.*

Thus, the sleep–wake cycle provides a starting point or framework for analyzing behavior (in other words, the output of the motor system) in a formal or systematic way. If we consider a typical 24-hour day (Fig. 7.1), about one-third of it is spent in an essentially continuous period of sleep—if we are lucky! Then we awake and begin to spend the day doing one thing after another until night comes, and we eventually start another cycle by going to sleep again.

From a scientific point of view, there are two important questions to ask and try to answer. First, what is the exact sequence of behaviors that are performed during the course of the day? We can simplify the problem by making the seemingly reasonable assumption that only one type of behavior can be performed at a time (a through i in Fig. 7.1), although one behavioral episode could be very brief. Second, why is a particular behavioral episode performed at any particular time? In other words, how are alternatives prioritized, and why does switching between behavioral episodes take place? From the structural perspective we are pursuing in this book, we would like to know the organization of neural circuits or networks that mediate switches between behavioral states, changes in priorities, and altered levels of arousal. These problems go to the very heart of nervous system function, and only the first can be answered with any degree of certainty at the present time. We can begin to approach the second by dividing each behavioral state (sleep and wakefulness) into components, and then dividing those components into fragments that lend themselves to neural systems analysis.

If the primary level of behavior analysis involves characterizing alternating periods of sleep and wakefulness, the secondary level involves characterizing the sequence of behaviors during wakefulness, as well as the fascinating repeating sequence of sleep stages that are displayed by all mammals. These stages were defined on the basis of human EEG (electroencephalograph or "brain wave") recordings in 1953 by Eugene Aserinsky and Nathaniel Kleitman, who scored a major conceptual breakthrough with this discovery. They showed that there is an alternation between what is called *rapid eye movement (REM) sleep*, when the cortical EEG is desynchronized (as in conscious attention) and vivid dreaming almost always occurs, and *deep sleep* (or non-REM sleep)—when dreaming is less vivid and perhaps less frequent. So there is an elaborate structure to the sleep state, and it is even more complex than first realized. Now we know based on characteristic patterns of the EEG that there is a sequence of four stages in a typical bout of deep sleep itself. In humans there is a startlingly regular cycle of alternating REM–deep sleep bouts during the course of an 8-hour period of sleep (Fig. 7.2). In the average adult, each bout of REM–deep sleep lasts about an hour to an hour and a

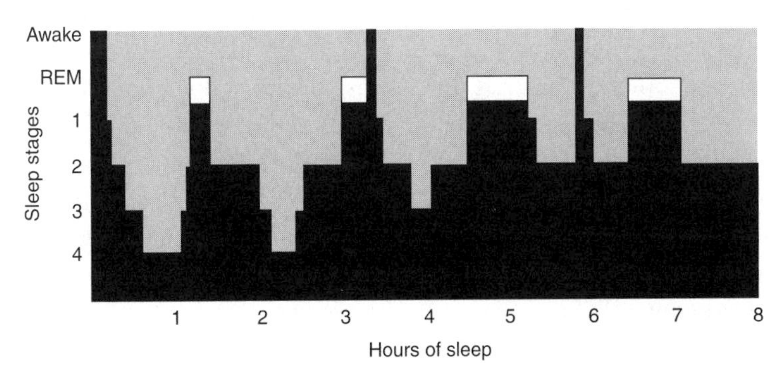

FIGURE 7.2 *The stages of sleep for a typical adult over an 8-hour period are shown in this graph. Adapted with permission from R.J. Berger, The sleep and dream cycle, in: A. Kales (ed.)* Sleep: Physiology and Pathology *(Lippincott: Philadelphia, 1969), p. 17–20.*

half, although bout length tends to decrease from childhood, through adulthood, and on into old age.

During wakefulness, a sequence of goal-oriented or motivated behaviors is displayed (a through h in Fig. 7.1). If we take a complete segment or episode of this behavior, it can be divided into three sequential phases (I, F, C in Fig. 7.1). First there is an initiation phase that triggers the search for a specific goal object or task. In the case of hunger, initiation stimuli (IS in Fig. 7.1) could include chemical signals related to low levels of blood glucose, the sight of an ad on TV for some tasty delight, or the smell of cooking dinner. Then there is the foraging phase when an exploratory strategy is used to find the goal object. This is also referred to as the procurement phase. Finally, there is the consummatory phase, when the goal object is used and associated with pleasant or unpleasant sensations and the behavioral segment comes to an end because of satiety mechanisms. Clearly, however, a behavior segment can be interrupted at almost any time by a different initiation stimulus if the stimulus is strong enough. Certain sensations associated with the consummatory phase play a critical role in positive and negative reinforcement of behavior. These are pleasant or unpleasant feedback signals that are associated with particular behaviors and help determine whether the particular behaviors are repeated or avoided in the future (Chapter 9).

To recapitulate, it would seem that during sleep the cognitive system is active while dreaming, and the sensory and somatomotor systems are somehow blocked. In contrast, during waking the cognitive and sensory systems modulate the output of the somatomotor system to produce behavior (see Fig. 5.5). The rest of this chapter delves further into the behavioral state system that controls the sleep–wake cycle, as well as the system that controls the level of arousal when awake. But first, it is illuminating to discuss a truly fascinating topic: circadian rhythms.

CIRCADIAN RHYTHMS *The Day–Night Cycle*

Throughout the entire span of evolution, life has been subjected to a day–night cycle where the length of the day varies in an extremely precise way with the yearly seasons. Thus it may not come as a total shock to learn that many animals have evolved endogenous clocks that produce a rhythm of activity with a period of about 24 hours (the definition of circadian). In mammals, a tiny, compact group of neurons is embedded in the hypothalamic visceromotor pattern generator network on either side of the third ventricle, and it is called the *suprachiasmatic nucleus*. Each nucleus produces a circadian rhythm of neuronal activity, and together they determine the pattern of the sleep–wake cycle and thus produce circadian rhythms in locomotor activity (walking around), eating and drinking, and a variety of more basic autonomic and endocrine responses. If a suprachiasmatic nucleus is removed from the brain and put into a dish, it can be kept alive for several days; during that time, it produces an endogenous circadian rhythm of activity.

In a very real sense, the suprachiasmatic nuclei are responsible for the fact that if people (or other animals) are placed in constant light or darkness (for weeks), they continue to show a sleep–wake cycle that is very similar to the one they display under normal conditions. There is, however, one curious twist: under constant lighting conditions, the time that people typically go to sleep is a half-hour later or so each day. The biological circadian clock has a period of about 24.5 hours (not the astronomical 24 hours), so that under constant lighting conditions it begins to "free-run." Under these conditions the circadian rhythm begins to drift in a predictable way, whereas under normal conditions the rhythm is synchronized to the day–night cycle by information from the eye. One of the more surprising findings of the 1970s was that the retina, and thus the optic nerve, has a direct neural input to the suprachiasmatic nucleus. This input provides information about environmental luminosity (very roughly, time of day, and even season of the year in terms of day length) to it.

Something incredible happens when the suprachiasmatic nuclei are lesioned—something that could never have been predicted before the experiments were actually conducted in the 1970s. Animals immediately lose their usual sleep–wake cycle (Fig. 7.3). Instead of

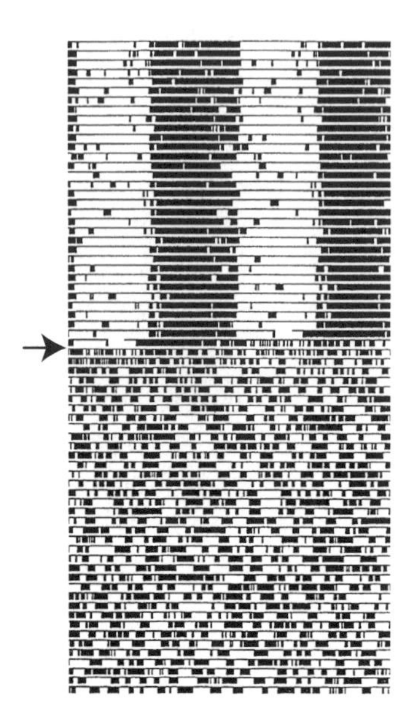

FIGURE 7.3 *The pattern of locomotor activity in a rat maintained on a 12-hour light–dark cycle, before* (above the arrow) *and after* (below the arrow) *bilateral lesions of the suprachiasmatic nuclei. The data is double-plotted for ease of interpretation: each line shows the preceding day and the new day. Note that in the intact animals, locomotor activity (dark bins) is confined to the period when it is dark (rats are nocturnal animals). In contrast, after suprachiasmatic nucleus lesions, the normal rhythm is abolished and locomotor activity is spread more or less evenly over the 24-hour period. Reproduced with permission from R.Y. Moore, Circadian timing, in: M.J. Zigmond, F.E. Bloom, S.C. Landis, J.L. Roberts, and L.R. Squire (eds.)* Fundamental Neuroscience *(Academic Press: San Diego, 1999), p. 1189–1191.*

sleeping more or less continuously for about 12 hours, and then staying awake for the next 12 hours (with an occasional nap), there are alternating periods of sleep and waking that last for an hour or so throughout the 24-hour period. There is still a sleep–wake cycle, but its periodicity is much shorter. As a matter of fact, it would seem that this is the natural period for the sleep–wake cycle and that the suprachiasmatic nuclei somehow impose a roughly 24-hour period on it—somehow convert an approximately 1-hour rhythm to an approximately 24-hour rhythm. Naturally, this leads to a complete reorganization of the animal's behavior. For example, eating and drinking in lesioned animals are distributed evenly across the 24-hour period instead of being concentrated in the continuous 12-hour period that the intact animal is awake each day. The extended periods of restlessness displayed by patients with Alzheimer's disease may be due in part to pathological lesions of their suprachiasmatic nuclei.

Quite recently there have been major breakthroughs in understanding the molecular basis of circadian rhythm generation in suprachiasmatic neurons and, surprisingly, in many other cells of the body. The key to this advance was the identification of fruitfly mutants with altered daily rhythms. Characterization of the affected genes has led to the identification of a group of genes whose protein products are involved in a complex program of gene expression changes with a circadian pattern—a pattern that leads to circadian changes of neuronal activity in the suprachiasmatic nucleus.

REPRODUCTIVE CYCLES

Many animals show a reproductive cycle that acts to maximize the production and survival of offspring. From the historical perspective of evolution, it is the single most important bodily function, because without it the species could not survive in nature. It would become extinct. In the stickleback fish that Nikolaas Tinbergen analyzed (see Fig. 6.10) the reproductive cycle is seasonal—it occurs once a year and is timed so that offspring are born during the spring. By way of contrast, women during their fertile years show an approximately lunar cycle that they go through 12 or 13 times a year, and female rats have an even shorter cycle that lasts only 4 or 5 days. In all three species, the peak of the reproductive cycle occurs around the time of ovulation, which is actually triggered in the brain—by neural activity in the GnRH neuroendocrine motoneuron pool in and near the rostral hypothalamus (see Figs. 6.11 and 6.13–6.15). These motoneurons cause a surge of anterior pituitary gonatotropic hormone secretions that, in turn, lead to ovulation and the secretion of gonadal sex steroid hormones.

It is these gonadal steroids (estrogens in females and androgens in males) that are responsible for activating reproductive behaviors (see Fig. 6.10). They do this by entering the brain, where they modify the gene expression that is related to neurotransmission in neural circuits that mediate reproductive behavior—in essence activating these circuits for specific functions like seeking out a mate and copulation. There is a sexually dimorphic circuit in the brain—one that is anatomically distinct like the genitalia and secondary sexual characteristics of the body—and its state is controlled by estrogens and androgens secreted into the blood from the gonads (see Chapter 10).

The basic idea behind these physiological and behavioral aspects of the reproductive cycle is shown beautifully in the rat, where they have been subjected to intense experimental analysis over the last century. Let's start with a very simple behavioral measure: the amount of walking around (locomotion or "activity") displayed by an animal. When this is measured in a prepubertal female rat, it is at a relatively low level, which is about the same every day (Fig. 7.4). However, when she reaches puberty, which is defined as the day she exhibits her first effective gonadotropin surge, the female rat runs around a lot and actually looks for a male to copulate with. She is in "heat" for about a day just before, during, and just after she has ovulated and is thus fertile. In the time before she had ever ovulated, and during the three days after ovulation, she displays no reproduc-

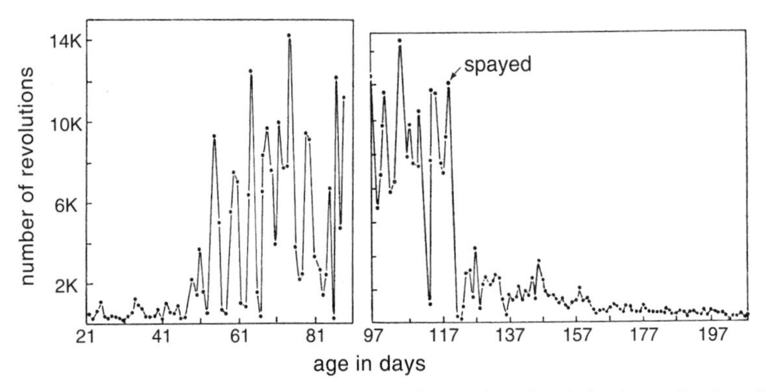

FIGURE 7.4 *This graph shows locomotor activity (number of revolutions) as a function of age in female rats. Note first that locomotor activity is quite low until about day 50, when the animal undergoes puberty and displays its first estrogen surge and ovulatory cycle. On the day of ovulation, the animal is in behavioral "heat" and displays a dramatic increase in locomotor activity, which is directed toward finding and attracting a mate. Then note that there is a major peak of locomotor activity every 4 days, the length of the estrous cycle in this animal. Finally, note that on day 119 the animal was ovariectomized (spayed). The estrous cycle is abolished immediately, along with the 4-day cycles of locomotor activity. The cycle can be restored by endogenous treatments with estrogen. Adapted from S.A. Barnett,* The Rat: A Study in Behavior *(Aldine: Chicago, 1963).*

tive interest in males, who, in fact, are actively avoided and vigorously defended against in the event of a sexual advance.

This 4- or 5-day cycle of ovulation and behavioral heat goes on relentlessly until menopause or until the gonads are removed (or their estrogens are blocked pharmacologically). The beauty of this model is that we know estrogens are responsible because the cycle can be restored in castrated animals by estrogen replacement. After this simple procedure, behavioral heat is displayed about 8 hours later, presumably the time it takes for estrogen to modify a specific (incompletely understood) gene expression pattern in the sexually dimorphic brain circuit that is required to produce the foraging and solicitous behaviors associated with being in heat (Fig. 7.4). By contrast, sexually mature male rats do not show a cycle of testosterone secretion. In essence, they are always interested in female rats, especially those in heat, who are probably secreting powerful pheromones. As in females, gonadal steroids contribute a great deal to "interest" in the opposite sex. In male rats, the sex drive is severely reduced by castration and is restored when they are treated with physiological replacement doses of testosterone.

The important generalization here is that the reproductive cycle plays a fundamental role in organizing behavior patterns, similar to (although not as dramatic as) that imposed by the sleep–wake cycle. Clearly, different behavioral states are associated with extremes of the reproductive cycle, and fluctuating levels of gonadal steroids play an important role in establishing these behavioral states. The reason for dealing with this topic here is that lesions of the suprachiasmatic nuclei abolish the normal reproductive cycle in rats, just as they abolish the normal sleep–wake cycle. Somehow, the circadian signal from the suprachiasmatic nucleus is converted to a 4- or 5-day signal destined for the gonadotropin surge generator. This conversion probably takes place in the anteroventral periventricular nucleus of the hypothalamus, which receives a neural input (a projection) from the suprachiasmatic nucleus and in turn projects to the GnRH neuroendocrine motoneuron pool that controls gonadotropin secretion from the pituitary gland. Lesions of this nucleus also abolish the reproductive cycle in rats.

SLEEP–WAKE CYCLES

We noted earlier in this chapter that total removal of the hypothalamic master circadian clock (the suprachiasmatic nuclei) produces alternating bouts of sleep and wakefulness that last about an hour to an hour and a half throughout the 24-hour period of each day (see Fig. 7.3). The neural mechanism (whether an individual cell group or a specialized network) responsible for generating this fundamental pattern of sleep and wakefulness is not known. Obviously, then, it is unclear how the suprachiasmatic nuclei convert this fundamental pattern into the typical situation where a person or animal sleeps more or less continuously for 8 to 12 hours a day and is awake the rest of the time.

Nevertheless, there is good experimental reason to suspect that the fundamental sleep–wake generator (or clock) resides in rostral regions of the pontine reticular formation (apparently in the oral/rostral part of the pontine reticular nucleus). Furthermore, this same general region is responsible for the generation of REM sleep, or at least several major features characteristic of REM sleep. For example, cholinergic neurons in the pedunculopontine tegmental nucleus project to the thalamus and appear to be critical for generating the ponto-geniculo-occipital (PGO) spikes characteristic of the EEG in REM sleep, and other pontine neurons with descending projections appear to be responsible for the muscle atonia (relaxation) that is characteristic of REM sleep. This occurs at least in part via an excitatory (glutamatergic) projection to the ventral medullary reticular formation, which, in turn, sends a descending inhibitory (at least partly glycinergic) projection to spinal motoneuron pools. It is intriguing to note that the bouts of sleep and wakefulness in animals with suprachiasmatic nucleus lesions are about an hour or so long (Fig. 7.3), as are bouts of REM and deep sleep (Fig. 7.2). These facts hint at the existence in the pons of a fundamental behavioral state rhythm generator with a period of about 90 to 120 minutes, which mediates both the underlying sleep–wake cycle and the REM–deep sleep cycle.

Extensive evidence exists, showing that certain specialized cell groups in the brainstem control the output of the fundamental behavioral state rhythm generator in the pons, control levels of arousal during wakefulness, and control the various stages of the sleep cycle itself. However, at this point in time, the extent to which these cell groups simply modulate behavioral state rhythm generators—as opposed to forming integral parts of the rhythm generator networks themselves—remains unclear. For the sake of convenience, then, we now consider a series of interesting cell groups in the brainstem that are characterized by the expression of a signature neurotransmitter (often a biogenic amine), by the elaboration of relatively widespread axonal projections, and by apparent involvement in modulating or controlling behavioral state.

MODULATING BEHAVIORAL STATE

In the early 1960s two young Swedish neuroscientists, Annica Dahlström and Kjell Fuxe, carried out a highly original series of neuroanatomical studies with a novel histochemical method that had just been developed by their mentors Bengt Falck, Arvid Carlsson, and Nils-Åke Hillarp for the demonstration of biogenic amine–containing neurons. Dahlström and Fuxe described in detail the overall organization of several neural systems that previously had been unsuspected and that were so unusual it took many years to convince the more skeptical neuroanatomists (who almost by trade are very conservative) of their reality.

One neural system contained noradrenaline, the same neurotransmitter used by the sympathetic division of the autonomic system (see Chapter 6). One noradrenergic cell group in particular stood out: the locus ceruleus. It soon became clear that neurons in this cell group of the pontine central gray send axons to innervate virtually the entire central nervous system—from the caudal end of the spinal cord, to the cerebellum and brainstem, to the entire cerebral cortex—in a seemingly diffuse, rather nonspecific way (Fig. 7.5). The locus ceruleus had been discovered almost 200 years earlier in the

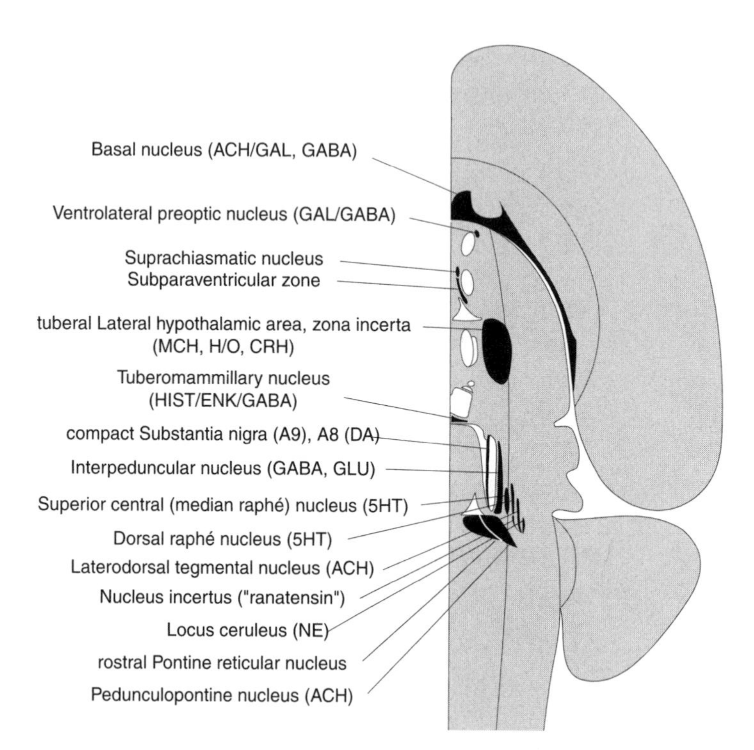

FIGURE 7.5 *Major cell groups associated with the behavioral state control system are shown in black on this flatmap of the rat central nervous system. The behavior control column (see Fig. 6.11) is shown in white. The compact part of the substantia nigra is sometimes referred to as the A9 group, whereas the ventral tegmental area contains the A8 group. Abbreviations for neurotransmitters: ACH, acetylcholine; CRH, corticotropin-releasing hormone; DA, dopamine; ENK, enkephalin; GABA, γ-amino butyric acid; GAL, galanin; GLU, glutamate; H/O, hypocretin/orexin; HIST, histamine; MCH, melanin-concentrating hormone; NE, norepinephrine; 5HT, serotonin.*

human brain by the French neuroanatomist Félix Vicq d'Azyr because it forms a tiny but obvious blue spot under the floor of the rostral fourth ventricle (Fig. 7.5). But in hindsight no one in the meantime had even a clue as to its true connections and neurochemistry—it almost seems to form a sympathetic ganglion in the brainstem. Needless to say, it has now been implicated in a broad range of functions. It seems safe to conclude that the locus ceruleus plays a role in the processing of all novel stimuli during the waking state and in switching between certain parts of the sleep–wake cycle. In rats the locus ceruleus is a virtually pure population of about 1600 noradrenergic neurons. Other central noradrenergic cell groups are all restricted to the hindbrain, and most of them appear to have more specialized functions than the locus ceruleus—functions that are especially related to the central autonomic control network.

A second system characterized by Dahlström and Fuxe uses serotonin as a neurotransmitter, and the neurons involved are found in the midbrain and hindbrain—mostly in and around a group of nuclei that had been described earlier, but had no known function. These are the raphé nuclei that lie in and near the midline of the adult brainstem. The two largest groups of serotonergic neurons are centered in the midbrain (Fig. 7.5), just rostral to the locus ceruleus, and are known as the *dorsal nucleus of the raphé* and the *superior central nucleus* (median nucleus of the raphé). These nuclei also have very widespread, seemingly diffuse, projections throughout much of the central nervous system, and they also play an important role in modulating or controlling various aspects of behavioral state.

A third system characterized by the two Swedes was centered in the ventral midbrain and uses dopamine as a neurotransmitter (Fig. 7.5). Unlike the noradrenergic and serotonergic systems just discussed, its projections are mostly if not wholly ascending. One specialization of this system is centered in the compact part of the substantia nigra, and its axons primarily innervate the dorsal striatum (part of the endbrain cerebral nuclei or basal ganglia). This dopaminergic pathway degenerates in Parkinson's disease, and treatment with the dopamine precursor L-dopa alleviates patient's tremors and inability to initiate behaviors, at least during early stages of the disease. The other specialization of the system is centered in the adjacent ventral tegmental area and the so-called retrorubral area, and it has more widespread ascending projections to the ventral striatum, prefrontal cortex, and hippocampal formation. Dahlström and Fuxe referred to the previously unknown dopamine neurons in these areas as the A8 and A10 cell groups, respectively. The dopaminergic compact part of the substantia nigra was called the A9 cell group.

The ventral tegmental area has been implicated in regulating levels of locomotor behavior (behavioral arousal) and in mechanisms of reward and positive reinforcement. The differential roles of dopaminergic and non-dopaminergic (probably GABAergic) neurons in the ventral tegmental area are not yet entirely clear. However, it seems possible that the former are involved in reward-related mechanisms, whereas the latter are involved in regulating locomotor behavior. This would be similar to the dopaminergic (compact) and non-dopaminergic (reticular) parts of the adjacent substantia nigra.

Thus, Dahlström and Fuxe laid out the basic neuroanatomy of three neurotransmitter-coded brainstem systems that are definitely related to controlling behavioral state in one way or another (although the details are still not resolved). Since then, a number of other nearby cell groups with similar functions have been identified (Fig. 7.5). For example, in the previous section of this chapter we discussed a cholinergic cell group centered in the pedunculopontine nucleus that is critically involved in modulating thalamocortical and other systems in relation to behavioral state, especially associated with REM sleep. Furthermore, a dorsally adjacent cholinergic cell group that lies next to the locus ceruleus in the pontine central gray has very widespread, seemingly diffuse, projections. This is the laterodorsal tegmental nucleus, and it has also been implicated clearly in the modulation of behavioral state. Finally, the obscure nucleus incertus should be mentioned: it also lies in the pontine central gray near the locus ceruleus, dorsal raphé, and laterodorsal tegmental nucleus. It is highly interconnected with two other brainstem midline nuclei (superior central and interpeduncular), and all three together project massively to forebrain systems associated with the prefrontal cortex and hippocampal formation. This midline trio of brainstem nuclei almost certainly plays a major role in behavior prioritization during the waking state.

As we move rostrally, we come to the lateral zone of the hypothalamus, which has often been thought of as the rostral end of the brainstem reticular formation. There are two especially interesting regions here. First we have the tuberomammillary nucleus, which is a thin collection of neurons that surrounds the mammillary body like a cradle. These neurons use histamine as one of their neurotransmitters (GABA is another one), and their axons have very widespread, seemingly diffuse, projections to most parts of the brain. They are the only neurons in the brain that synthesize histamine, and the drowsiness associated with antihistamines is thought to result from interfering with their normal function.

The second interesting feature of the lateral hypothalamus in this context is the presence of three separate, though intermixed, neuronal populations at the level of the ventromedial and premammillary nuclei (Fig. 7.5). All three populations appear to have very widespread projections to many parts of the central nervous system, including the brainstem, spinal cord, and cerebral cortex. One cell group uses the peptide melanin-concentrating hormone (MCH) as one of its neurotransmitters, another uses the peptides hypocretin/orexin and dynorphin as several of its neurotransmitters, and the other uses corticotropin-releasing hormone (CRH) as one of its neurotransmitters under certain conditions (as in anorexia). It has long been thought that the lateral hypothalamus plays an important role in modulating behavioral state, but this was shown dramatically with the discovery that mutations in the hypocretin/orexin gene, or in the gene for its receptor, cause narcolepsy in which both animals and humans have difficulty staying awake. This is the only population of neurons in the brain that expresses the hypocretin/orexin gene.

Finally we come to the cerebral or basal nuclei of the endbrain. Here we encounter a population of cholinergic neurons that has a topographically organized, though widely overlapping, projection to the entire cerebral cortical mantle. In the 1960s it was associated with the primate basal nucleus of Meynert, which had been discovered a century earlier. The cholinergic neurons are distributed irregularly, sometimes clumped and sometimes widely spaced, throughout the ventral pallidum and medial septal-diagonal band complex, and then become more sparse in the dorsal pallidum and even in parts of the striatum. The precise function of these neurons is unclear, although they degenerate in Alzheimer's disease and are thought to play a role in learning and memory mechanisms and in the modulation of behavioral state. It is now known that the so-called basal forebrain projection to cortex involves other cell types as well.

In summary, anatomically and neurotransmitter distinct neuronal cell groups stretching from the pons and midbrain through the hypothalamus to the cerebral nuclei play a critical role in modulating behavioral state. It seems clear that each of the cell groups has a specialized function—although exactly what those functions are remains to be determined—and that they are highly interconnected to form an exceedingly complex neural network. Presumably, specific patterns of activity within the network determine not only behavioral state but also levels of arousal within particular states. This network can be thought of as separate from the purely sensory and motor systems, as well as from the cognitive system of the cerebral hemispheres (see Fig. 5.5). In a gross anatomical sense, the neural network is closely associated with the behavior control column discussed in Chapter 6 (Figs. 6.11 and 7.5).

READINGS FOR CHAPTER 7

Fuxe, K., Hökfelt, T., Jonsson, G., and Ungerstedt, U. Fluorescence microscopy in neuroanatomy. In: W.J.H. Nauta and S.O.E. Ebbeson (eds.) *Contemporary Research Methods in Neuroanatomy*. Springer-Verlag: New York, 1970, pp. 275–314.

Goto, M., Swanson, L.W., and Canteras, N.S. Connections of the nucleus incertus. *J. Comp. Neurol.* 438:86–122, 2001.

Hobson, J. A. *The Dreaming Brain*. Basic Books: New York, 1988. This is a good overview of the topic from many perspectives.

Jones, B.E. The neural basis of consciousness across the sleep-waking cycle. *Adv. Neurol.* 77:75–94, 1998.

Jouvet, M. *The Paradox of Sleep: The Story of Dreaming*. MIT Press: Cambridge, Mass., 1999. Thoughts from a pioneer in the neurobiology of sleep.

Klein, D.C., Moore, R.Y., and Reppert, S.M. (eds.) *Suprachiasmatic Nucleus: The Mind's Clock*. Oxford University Press: New York, 1991.

Moore, R.Y. Circadian rhythms: basic neurobiology and clinical applications. *Ann. Rev. Med.* 48:253–266, 1997.

Rechtschaffen, A., and Siegel, J. Sleep and dreaming. In: E.R. Kandel, J.H. Schwartz, and T.M. Jessell (eds.) *Principles of Neuroscience*, fourth edition. McGraw-Hill: New York, 2000, pp. 936–947.

Rodrigo-Angulo, M.L., Rodriquez-Veiga, E., and Reinoso-Suárez, F. Serotonergic connections to the ventral oral pontine reticular nucleus: implications in paradoxical sleep modulation. *J. Comp. Neurol.* 418:93–105, 2000.

Saper, C.B. Diffuse cortical projection systems: anatomical organization and role in cortical function. In: *Handbook of Physiology: The Nervous System.* Waverly Press: Baltimore, 1987, pp. 169–210.

Saper, C.B., Chou, T.C., and Scammell, T.E. The sleep switch: hypothalamic control of sleep and wakefulness. *Trends Neurosci.* 24:726–731, 2001.

8

The Cognitive System

Thinking and Voluntary Control of Behavior

Thus there remains for exposition what it is that initiates voluntary movements, sensations, and that Reigning Soul, by which we imagine, meditate, and remember. To this task the present book is devoted.

—ANDREAS VESALIUS (1543)

I divide the functions of the brain into two classes: *vis.*, affective and intellectual; and, in harmony with this physiological division, I recognize two kinds of cerebral parts. The anterior pyramidal bodies [pyramids] I consider the rudiments of such as belong to the intellectual operations; and the other bundles of the medulla oblongata . . . which run across the annular protuberance [pons] to communicate with many of the cerebral masses, as the roots of those that pertain to the affective manifestations. . . . This separation into two systems of parts is very evident from the medulla oblongata upwards, as far as the pretended optic thalami [diencephalon] and striated bodies [basal ganglia] in man and the mammalia.

—JOHANN SPURZHEIM (1826)

The complexity of the nervous system is so great, its various association systems and cell masses so numerous, complex, and challenging, that understanding will forever lie beyond our most committed efforts.

—SANTIAGO RAMÓN Y CAJAL (1909)

The cerebral cortex is the crowning glory of evolution. It is the part of the nervous system that is responsible for thinking. Quadriplegics, who tragically have had their spinal cord disconnected from their brain, can think just fine, and so can people who have been born without a cerebellum. But extensive damage to the cerebral cortex profoundly interferes with cognition. The cerebral cortex is the organ of thought. Can the organ of thought ever understand itself? Will we ever understand the physical basis of thought? Can we ever discover the fundamental organizing principles of cerebral cortex circuitry? What is the biology of consciousness? If nothing else—how far have we come in our attempts to understand the brain substrates of thinking?

It was not so long ago that even trying to answer this question was dangerous. Franz Joseph Gall was the first physician–neuroscientist to argue that thinking takes place in the gray matter of the cerebral cortex and that different aspects of cognition are elaborated in different regions of the gray matter. However, in 1802 the German emperor Francis the First forbid Gall to lecture publicly or privately about this topic, on the grounds that it was materialistic and thus antireligious. Three years later Gall left his native land forever—banished—and began traveling through Europe, finally settling in Paris in 1807. This is how his monumental four-volume *Anatomie et physiologie du système nerveux* . . . , which was written in collaboration with his younger colleague Johann Caspar Spurzheim, came to be published between 1810 and 1819 in Paris rather than in Vienna.

Gall and Spurzheim also postulated that an unusually large particular region of gray matter and its corresponding overdeveloped mental function would be reflected in an expansion (or bump) in the overlying region of the skull. This assumption spawned the pseudoscience of phrenology, whose practitioners, led initially by Spurzheim himself, tried to determine people's mental gifts and deficiencies by analyzing the exact shape of the skull. But even in contemporary France there was widespread opposition to the idea of functional localization within the cerebral cortex itself. This hesitation was based largely on the experiments of the great physiologist Marie-Jean-Pierre

Flourens. Flourens used the ablation method in the 1820s to conclude that although the cerebral hemispheres are indeed the seat of intelligence and sensation, they are not functionally partitioned. He interpreted his results to indicate that, instead, the special senses and intellectual faculties are represented or distributed throughout the hemispheres. This view was not significantly eroded until the 1860s when the pioneering work of Paul Broca, Gustav Fritsch, Eduard Hitzig, and Hermann Munk began to establish cerebral localization for speech, motor control, and the various sensory modalities. Today, cognitive neuroscientists are using functional magnetic resonance imaging (fMRI) studies on living human brains to measure individual differences in cortical localization for every conceivable psychological factor. The basic principles underlying phrenology—cortical localization of function, as well as individual differences in cortical specialization—are being exploited as never before.

CEREBRAL CORTEX REGIONALIZATION

In Chapter 4 we found that each cerebral hemisphere (also known as cerebrum, endbrain, or telencephalon) may be divided into two very different parts. One part lies more dorsally and is a layered, sheet-like tissue known as the cerebral cortex or the pallium. The other part lies more ventrally and does not have a laminated appearance. It has been referred to variously as the basal ganglia, basal nuclei, or cerebral nuclei. The cerebral cortex is undoubtedly the best place to start because there is widespread agreement about its fundamental organization, which is relatively straightforward.

The basic division between cerebral cortex and nuclei becomes evident in the early embryo at the five-vesicle stage of the neural tube illustrated in Figure 4.10. Here the cerebral hemisphere has the simple shape of a contact lens or slightly flattened hemisphere, with the transition between cortex region and basal nuclei region indicated by a shallow internal groove, the corticobasal (or corticostriatal) sulcus (Fig. 8.1). As the mammalian cerebral cortex develops further, it evaginates or balloons out tremendously (Fig. 8.2) and eventually be-

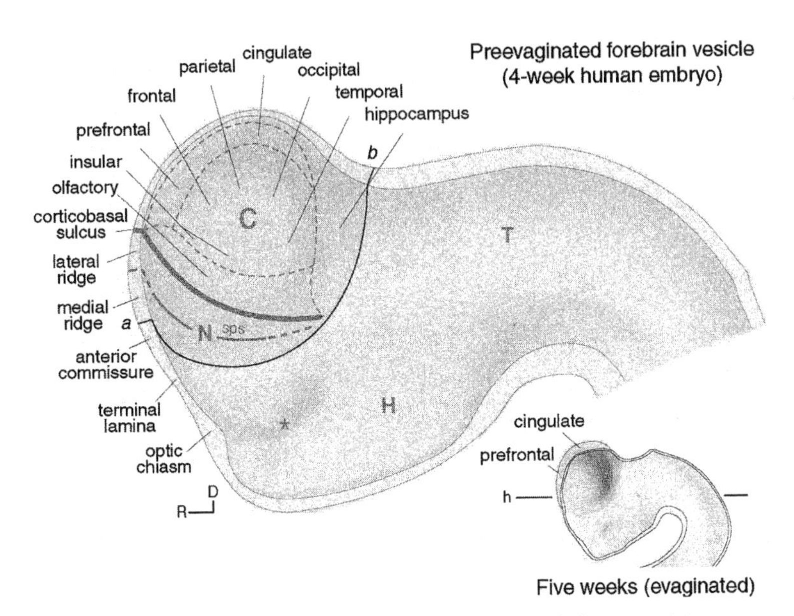

FIGURE 8.1 *The large drawing shows the right forebrain vesicle of the 4-week human embryo, with a fate map superimposed on the endbrain or cerebral hemisphere. Note that the cerebrum will be divided into cortex (C) and nuclei (N). This is an instructive stage because the hemisphere has not yet begun to evaginate (see smaller drawing where it has just begun and Fig. 8.2 where it is more advanced), so a qualitative fate map of cortical regionalization is easy to plot.* Key: *a, b, ends of a line that separates endbrain and interbrain components of the forebrain vesicle, the prospective foramen of Monro; D, R, dorsal and rostral; h, a horizontal line indicating the approximate plane in Fig. 8.2; H, hypothalamus; sps, striatopallidal sulcus; T, thalamus, *, optic vesicle. Adapted with permission of Elsevier Science from L.W. Swanson, Cerebral hemisphere regulation of motivated behavior,* Brain Res., 2000, vol. 886, p. 117.

comes corrugated or folded to a greater or lesser extent in different species. The development of cortical folding into gyri that are separated by sulci (and deeper fissures) takes place because the skull eventually limits unfettered expansion of the cortical sheet; as a result, much more cortical surface area may be squeezed into a limited volume. This folding of the cortical sheet is obvious in the brains in our frontispiece and in Figure 8.3. It is responsible for the fact that about two-thirds of the human cortical sheet (which has a surface area of

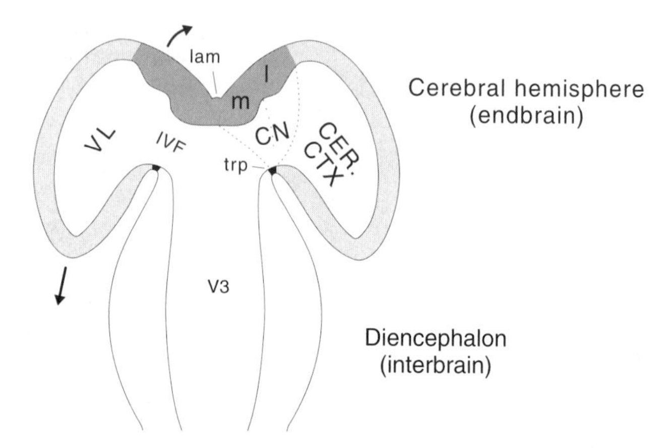

FIGURE 8.2 *This is a schematic horizontal section through the forebrain vesicle as the cerebral hemisphere is beginning rapidly to evaginate and expand (in directions shown by arrows; compare with Fig. 8.1).* Key: *CER. CTX, cerebral cortex; CN, cerebral nuclei; IVF, interventricular foramen (of Monro); l, lateral ventricular ridge; lam, terminal lamina; m, medial ventricular ridge; trp, telencephalic roof plate; V3, third ventricle; VL, lateral ventricle. Adapted with permission from L.W. Swanson,* Brain Maps: Structure of the Rat Brain *(Elsevier Science: Amsterdam, 1992, p. 33).*

about one square foot per hemisphere) is buried below the outer, readily visible surface of the hemisphere next to the skull.

The point of this discussion is that the adult cerebral cortex is a sheet that, at least in principle, can be flattened. Topologically it can be thought of as a flat sheet that is attached along its lower or ventral border to the cerebral nuclei, a relationship that is crystal clear in the early embryo (Fig. 8.1). This basic relationship does not change as the embryonic cerebrum develops into the adult. So it is possible to divide up the cortical sheet into structurally and functionally different regions, like a map of Europe can be divided into a number of different countries. This division or regionalization scheme has emerged from a vast amount of research in the century and a half since Broca, Fritsch, Hitzig, and Munk began the work mentioned here. And, like the map of Europe over the same period of time, the boundaries between regions have changed considerably and have

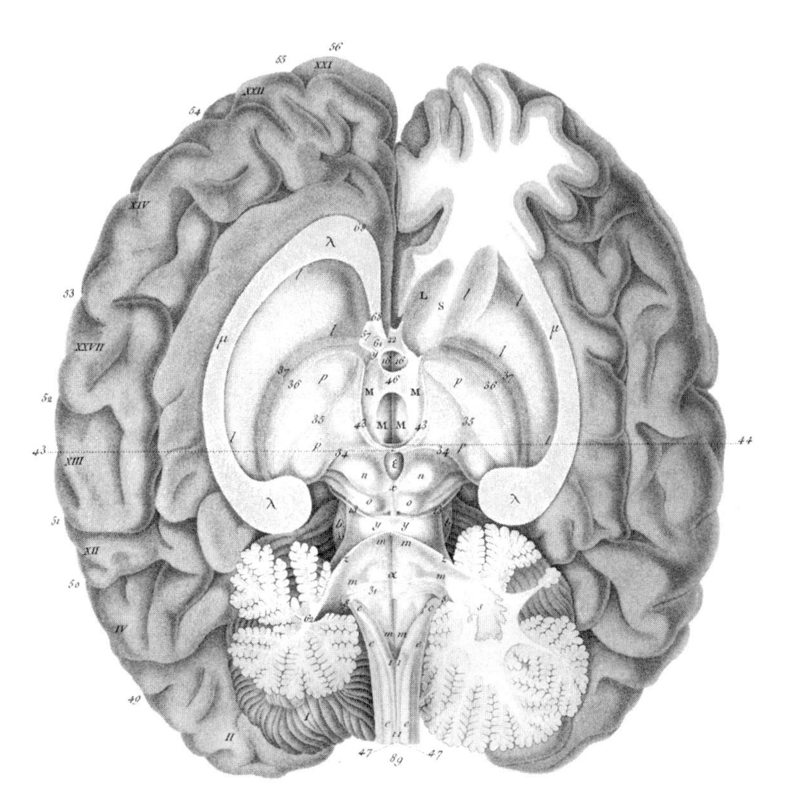

FIGURE 8.3 *In this dissection the human brain has been partly bisected from above and opened up like a book. Therefore one is viewing the medial surfaces of the cerebral hemispheres and the dorsal surface of the medulla. In addition, the frontal pole of the right cerebral hemisphere has been sliced off to reveal the characteristic pattern of gray matter and white matter associated with the cerebral cortex. The deep cerebellar gray (the dentate nucleus) is nicely illustrated, embedded within the cerebellar white matter (arbor vitae) on the right side. From F.J. Gall, and J.C. Spurzheim,* Anatomie et physiologie du système nerveux en général et du cerveau in particulier. *(Schoell: Paris, 1810–1819).*

been subject to different interpretations by different parties. As more is learned about cerebral organization, the map is bound to evolve.

The first hint of cortical regionalization came from the research of an Italian medical student, Francesco Gennari, who in 1776 noted the presence of a distinct layer of white matter in caudal regions of the freshly sliced human cerebral cortex. As we shall soon see, it became clear many years later that the famous "stripe of Gennari" is a characteristic feature of one particular human cortical area—the primary visual area of the occipital lobe, and more specifically layer 4 of this area, which is also known as the striate area because of the characteristic stripe. At about the same time, the great biologist Albrecht von Haller began referring to large expanses of cortex with reference to the bones of the skull overlying them: hence, the frontal, parietal, occipital, and temporal regions or lobes. Then, beginning with Theodor Meynert in the 1860s, neuroanatomists began producing a succession of maps that more or less systematically describe structural regionalization throughout the cortical sheet or mantle based on histological criteria. The most famous and enduring cortical maps were generated by Korbinian Brodmann, whose 1909 book on the topic is an intellectual tour de force and a great classic in neuroscience. Based on how neuronal cell bodies tend to form layers in different regions of the cortical mantle, Brodmann recognized about 50 distinct cortical areas in mammals, which he numbered in an arbitrary way (Fig. 8.4). The topological relations of these areas are illustrated on a cortical flatmap in Figure 8.5.

Whether acknowledged or not, all critical work done in the last century on cerebral cortical structure and function is derived from the maps produced by Brodmann. What has been done in the meantime is to assign functional significance to many of the areas, and often to parcel the original areas even further. Although there is a great deal of controversy about details, broadly speaking it would appear that the cortical mantle is divided into motor areas, unimodal sensory areas, and association areas where more than one sensory and or motor modality converge. In a particularly insightful 1970 article, Edward Jones and Thomas Powell suggested that information from each sensory modality follows a similar progression of connections through the cortex, at first separately and then eventually converging in polymodal association areas.

First, sensory information reaches a primary unimodal sensory area (illustrated by dark shading in Fig. 8.5). Then it is relayed to one or more adjacent, corresponding unimodal sensory association

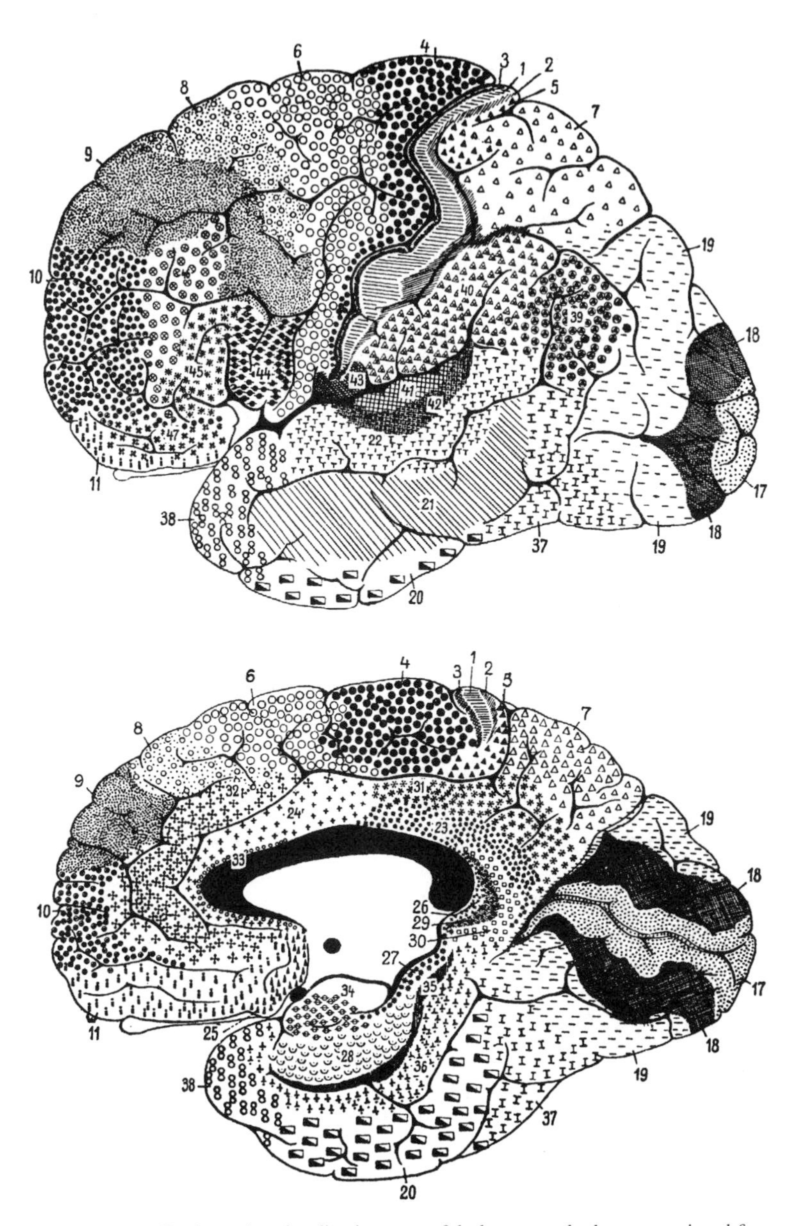

FIGURE 8.4 *Brodmann's regionalization maps of the human cerebral cortex as viewed from the lateral* (top: *left hemisphere*) *and medial* (bottom: *right hemisphere*) *aspects. Each cortical area is indicated by a different pattern of symbols and a different number. From K. Brodmann,* Vergleichende Localisationslehre der Grosshirnrinde in ihren Prinzipien dargestellt auf Grund des Zellenbaues *(Barth: Leipzig, 1909).*

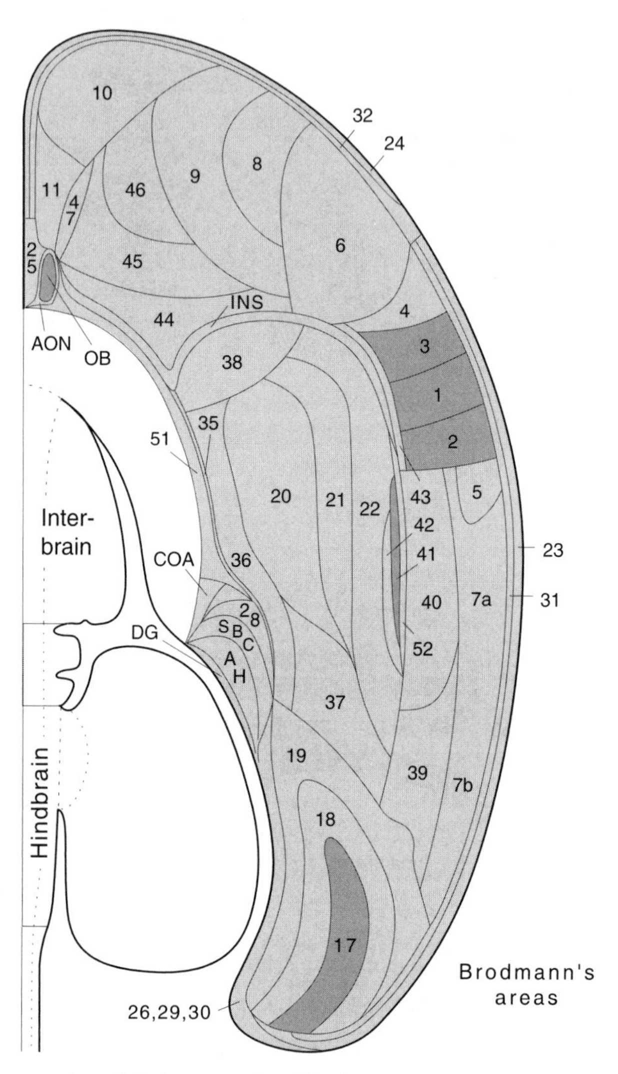

FIGURE 8.5 *A topological representation of Brodmann's regionalization of the cerebral cortex into areas is shown on this flatmap of the human brain (see Fig. 8.4). Primary sensory cortical areas are indicated in darker gray (1–3, somatosensory; 17, visual; 41, auditory; OB, olfactory bulb).* Key: *AH, Ammon's horn; AON, anterior olfactory area, COA, cortical amygdalar area; DG, dentate gyrus; INS, insular area; SBC, subicular complex. Adapted with permission of Elsevier Science from L.W. Swanson, Mapping the human brain: past, present, and future,* Trends Neurosci., *1995, vol. 18, poster accompanying p. 471.*

areas, where more complex processing of the particular mode of sensory information takes place—as well as to one or more motor-related areas in the frontal lobe. Then, from each unimodal association area, there are projections to adjacent association areas and to other motor-related regions in the frontal lobe as well. Eventually, unimodal association areas project to polymodal association areas, where information from two or more sensory modalities converges. In turn, polymodal association areas project to motor areas, as well as back to unimodal sensory areas. For example, Brodmann's area 17 would be considered the primary visual cortex; his areas 18 and 19 would be considered unimodal visual association cortex; the inferior temporal cortex (area 20) and the hippocampal formation would be considered polymodal association cortex; and areas 4, 6, and 8 would be considered primary and supplementary motor cortical areas.

While it has been found very useful to think about cortical function in this way—by following the course of a particular sensory modality after it reaches the primary cortical areas—the actual situation is much more complex than this. Each cortical area receives information from a specific nucleus or set of nuclei in the thalamus, as well as a variety of inputs from the brainstem behavioral state control system (see Chapter 7 and Fig. 7.5). In a real sense, then, each cortical area is under constant parallel control by ascending inputs, even though serial processing of particular sensory inputs may be taking place simultaneously through particular intracortical pathways. The actual dynamics of information processing within the network of connections between the various cortical areas is far from understood.

At a very basic level, though, the connections forming this intracortical network may be divided into two classes: association and commissural. Association connections are established between cortical areas within the same hemisphere, whereas commissural connections are formed between cortical areas in the right and left hemispheres. Commissural pathways crossing the midline and thus interconnecting the two hemispheres include the anterior commissure, the great cerebral commissure (corpus callosum), and the hippocampal commissures. Because of their complexity, the actual organization of intracerebral connections may well lie beyond the limits of human comprehension.

CORTICAL CELLULAR ORGANIZATION

Cortical lamination patterns have been referred to again and again. What are they? The feature that Brodmann and many others have exploited is the distribution of neuronal cell bodies within the cortical sheet. In the 1890s Franz Nissl developed a stain for showing very clearly the location, size, and shape of neuronal somata in histological sections of the brain, and his method has become perhaps the most widely used in neuroanatomy for its simplicity, reliability, and utility. This is the method that Brodmann applied to the cerebral cortex of a wide variety of mammals. In essence, he recognized about 50 different or distinct lamination patterns in the cerebral cortex. At the most general level, he divided them into two classes. One class passed through a clear six-layered stage during development, and he referred to it as the *homogenetic cortex*. The other class did not pass through a six-layered stage during development, and he referred to it as the *heterogenetic cortex*. A few years later, in 1919, Oskar and Cécile Vogt applied the terms *isocortex* and *allocortex* to the homogenetic and heterogenetic cortices, respectively, and the Vogts' terms are preferred today. Other still popular terms are *neocortex* (for isocortex), and *paleo-* and *archicortex* (together for allocortex), but they are based on unfounded evolutionary arguments from around the end of the nineteenth century.

Perhaps the best example of different lamination patterns in the isocortex involves the adjacent areas 17 and 18—the primary visual and unimodal visual association areas, respectively. Figure 8.6 is a photomicrograph, taken from Brodmann's work, of a Nissl-stained tissue section cut perpendicular to the surface of the human cortex, with the border between areas 17 and 18 running down the middle. Note the six classical cell layers of isocortex in both areas, then look

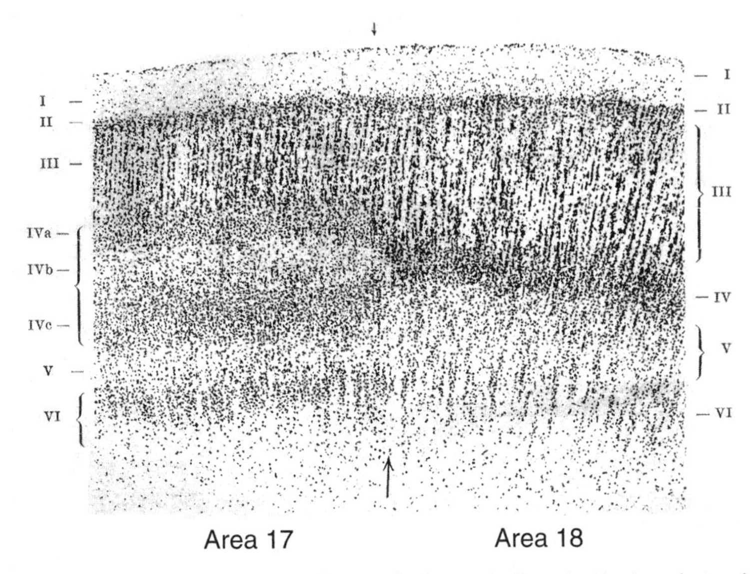

FIGURE 8.6 *The cytoarchitecture of human visual cortex is shown in this photomicrograph of a Nissl-stained tissue section. The large arrow near the bottom indicates the border between area 17* (left) *and area 18* (right). *Layers are identified by the Roman numerals at either end. The large arrow is in the deep white matter of the cortical mantle. The outer surface of the cortex is at the top (near the small arrow). From K. Brodmann,* Vergleichende Localisationslehre der Grosshirnrinde in ihren Prinzipien dargestellt auf Grund des Zellenbaues *(Barth: Leipzig, 1909).*

at layer 4 (IV). Throughout the isocortex, layer 4 is characterized by small neurons, and for this reason it was long ago named the *granular* or *granule cell layer.* You can see that in area 18 (on the right), layer 4 is relatively uniform from superficial (surface of the cortex at the outer edge of layer 1) to deep (toward the deep white matter below layer 6), whereas in area 17 (on the left) layer 4 is very clearly split into three sublayers, a, b, and c. Clearly, this expansion and differentiation of layer 4 has profound effects on the thickness of adjacent layers 3 and 5 in area 17. There is no doubt that the laminar distribution of neurons in areas 17 and 18 is quite different, although the differences appear to be quantitative variations on a six-layered scheme.

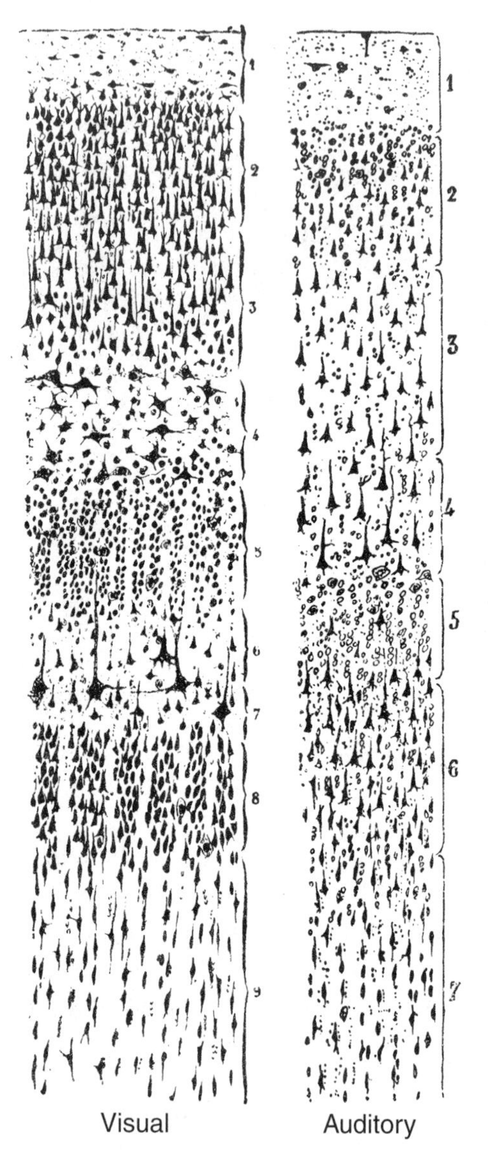

FIGURE 8.7 *The cytoarchitecture of human visual and auditory cortex is compared in these drawings of Nissl-stained sections by Cajal. Note that Brodmann later used a different numbering scheme for the lamination pattern (see Fig. 8.6, right half, for visual cortex). From S.R. Cajal,* Histologie du système nerveux de l'homme et des vertébrés, *vol. 2 (Maloine: Paris, 1911).*

There are also clear, though perhaps more subtle, differences between the various primary sensory cortical areas. Figure 8.7 illustrates Cajal's interpretation of lamination differences between visual cortex and auditory cortex in the adult male human brain. Here Cajal has chosen to illustrate his findings with drawings of Nissl-stained sections rather than with photographs, which he virtually never used. Disregarding Cajal's alternative lamination numbering scheme (his layer 5 corresponds to Brodmann's layer 4), it is obvious that there are differences throughout the thickness of each cortical area. One can only assume that these differences are somehow responsible in part for the qualitative differences between visual and auditory sensations.

Over the years structural neuroscientists have attempted to determine how many different cortical areas they can distinguish on the basis of such differences in lamination patterns. Unfortunately, many of the distinctions that have been made are considerably more subtle than those illustrated here, and some authors have argued that in certain regions of association cortex gradients rather than clear borders are found between adjacent areas. Overall, the number of cortical areas distinguished by various investigators ranges between about 20 and 400.

Estimates of the total number of neurons in the human cerebral cortex on both sides of the brain range from 3 to 14 billion. This is a lot of neurons by any account, and there are on the order of 10 times as many glial cells. Fortunately, it seems reasonable to assume that all of these neurons fall into two broad classes: pyramidal neurons that have long projections, and stellate neurons that have local circuit connections within a particular cortical area. This distinction was originally made in a brief, three-page report by Camillo Golgi in 1873. Here, in what is probably line for line the most important paper in the history of neuroscience, the young Italian physician reported a radically new method for staining individual neurons in their entirety. Based on application of this method, he discovered that, contrary to conventional wisdom, dendrites do not anastomose with one another; he described axon collaterals accurately for the first time; and he divided all neurons (including those in cerebral cortex) into projection and local circuit classes.

In the isocortex there are very few neurons in layer 1; layers 2 and 3 are characterized by relatively small pyramidal neurons (along with local circuit neurons), layer 4 consists almost entirely of local circuit (granule) cells, and layers 5 and 6 are characterized by larger pyramidal neurons (along with local circuit neurons) (Fig. 8.8). Interestingly, the smallest pyramidal neurons tend to be localized to layer 2, and they typically generate association projections to other

FIGURE 8.8 *This drawing indicates some of the major neuronal cell types (1–17) found in various layers (I–VIb) of the cerebral isocortex, along with the morphology of certain major classes of axons that enter the cortex to end (a–f). The cortical layers are identified with the Nissl stain, and the neuronal cell types and afferent fibers are drawn from Golgi preparations. Reproduced with permission from a drawing by Raphael Lorente de Nó in J.F. Fulton,* Physiology of the Nervous System *(Oxford: London, 1938, p. 302)*

cortical areas within the same hemisphere. Layer 3 pyramidal neurons tend to be somewhat larger, and they typically generate commissural projections to the opposite hemisphere, as well as association projections to the same hemisphere. The largest pyramidal neurons are found in layers 5 and 6, and they are responsible for generating most of the descending cortical projections to the cerebral nuclei, brainstem (including the thalamus), and spinal cord.

This arrangement suggests that the lamination patterns characteristic of the cerebral cortex are due to the differential distribution of pyramidal neuron subpopulations characterized by different projection terminal fields (for example, association projections versus commissural projections versus descending projections to the cerebral nuclei, thalamus, or brainstem and spinal cord). Modern pathway tracing experiments bear this assumption out and show that each cortical area subjected to careful analysis has a distinct pattern of projections to other parts of the brain; these projections are presumably generated by different subpopulations of pyramidal cells, which are reflected in more or less obvious lamination patterns.

This arrangement also suggests a fundamental organization of isocortex into three "super layers": supragranular, granular (layer 4), and infragranular. Layer 4 is characterized by a dense input from the thalamus (Fig. 8.8), and much of the local circuit output of layer 4 stellate neurons is directed toward the supragranular layers. The supragranular layers of relatively small pyramidal neurons generate primarily intracortical projections. In essence, the supragranular layers generate the immensely complex network of connections between cortical areas, and Cajal was perhaps the first to emphasize the possibility that this network is primarily responsible for thinking, learning, and memory. The supragranular layers also provide a major input to the infragranular layers of relatively large pyramidal neurons that generate most of the output of cerebral cortex to other parts of the brain. In other words, the infragranular layers are essentially the "motor" part of the cerebral cortex. According to this scheme for the isocortex, infragranular layers execute the cognitive computations elaborated in supragranular layers.

CORTICAL PROJECTIONS

As just mentioned, the majority of descending projections from the isocortex arise from pyramidal neurons in the infragranular layers. Broadly speaking, these projections seem to arise from three classes of pyramidal neurons. One class dominates layer 6 and projects to the thalamus. Another class dominates in superficial layer 5, and it is characterized by an input to the striatal component of the cerebral nuclei. The other class dominates in deep layer 5, and it projects to the brainstem and spinal cord (but apparently not heavily to the striatum). As a whole, these descending cortical projections innervate primarily the motor and sensory systems as defined in Chapters 6 and 9, respectively (see also Fig. 5.5).

It should come as no surprise that the organization of descending cortical projections, which provide the cognitive influence on behavior, are extensive and exceedingly complex. This is a topic that is far beyond the scope of the present discussion and can be delved into more deeply in the references listed at the end of the chapter. However, it is important to appreciate that all of these cortical projections are topographically organized and fundamentally based on the regional map indicated in Figures 8.4 and 8.5. That is one reason the importance of understanding the principles of cortical regionalization cannot be overemphasized. For example, virtually the entire cortical mantle shares bidirectional connections with the thalamus and also sends a topographically organized projection to the striatum.

The cortico-striatal projection leads us to consider the other half of the cerebral hemisphere—the cerebral or basal nuclei.

THE CEREBRAL NUCLEI

As we saw earlier (Fig. 8.1), the cerebral cortex and the cerebral or basal nuclei are completely distinct structures in the endbrain vesicle of the very young embryo. However, because cerebral nuclei undergo relatively much more growth than cerebral cortex, they bulge inward and come to lie adjacent to parts of the cortex in the maturing (Fig. 8.9) and adult (Fig. 8.10) brain. There is a great deal of confusion and disagreement about exactly what constitutes the cerebral nuclei and, in fact, about how to group the various components of the cerebral hemispheres. In dealing with the hemispheres, one encounters terms including limbic system, septum, amygdala, extended amygdala, rhinencephalon, corpus striatum, dorsal and ventral striatum and pallidum, neocortex, and so on. As mentioned earlier in this chapter, one way around this miasma is to adopt the view that the cerebral hemispheres consist simply of cortex and nuclei, with the latter divided into striatum and pallidum. If one simply begins with the regional map of cerebral cortex and considers the function and connections of its various areas, many of the terms just listed can be seen as unnecessary or even arbitrary and confusing.

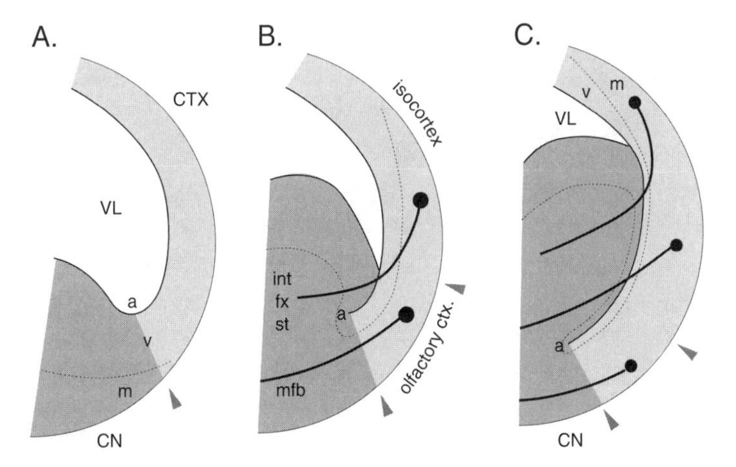

FIGURE 8.9 *The differential growth of the cerebral nuclei (CN; darker gray) as compared to the cerebral cortex (CTX; lighter gray) during cerebral hemisphere embryogenesis is shown in these schematic drawings (from early, A, to later, C).* Key: *a, ventral angle of lateral ventricle; fx, fornix; int, internal capsule; m, mantle layer of neural tube; mfb, medial forebrain bundle; st, stria terminalis; v, ventricular layer of neural tube. Adapted with permission of Elsevier Science from L.W. Swanson, and G.D. Petrovich,* Trends Neurosci., *1998, vol. 28, p. 325.*

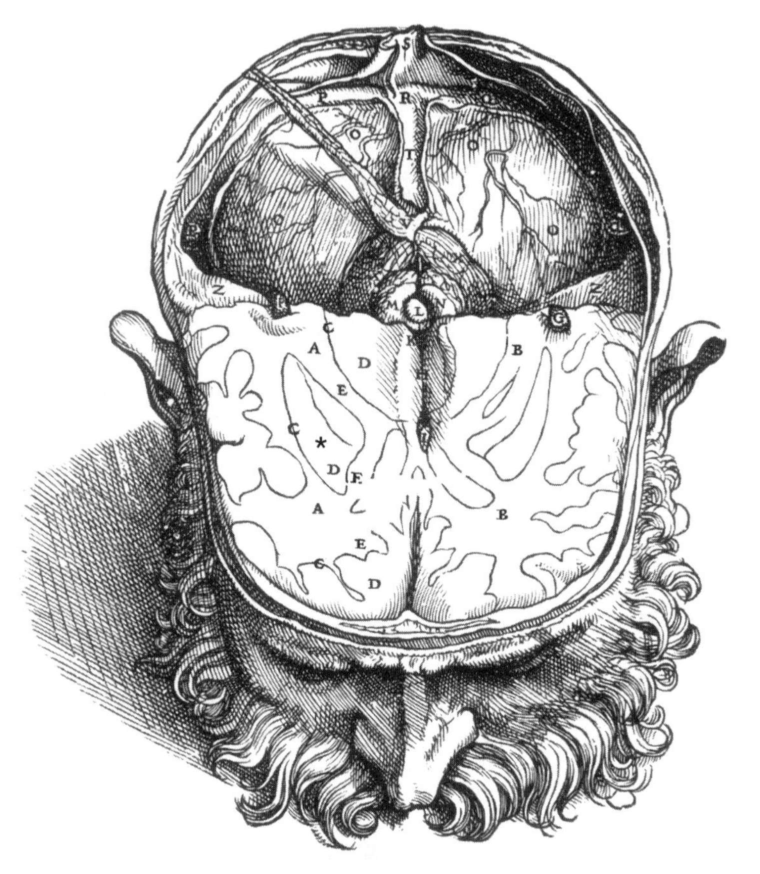

FIGURE 8.10 *This drawing of the partly dissected head and brain is from Vesalius's* Fabrica *(1543). The caudal halves of the cerebral hemispheres have been removed to reveal the pineal gland (L), tectum (M, N), and cerebellum. A horizontal slice through the remaining cerebral hemispheres clearly reveals the pattern of gray and white matter. Note that on the left-hand side (the right hemisphere), the cerebral nuclei or basal ganglia are clearly outlined (indicated by an asterisk, *, which has been added). It is one of a magnificent series of 15 drawings that illustrate a complete dissection of the brain. Compare with Figure 5.1.*

This simple view of cerebral hemisphere regionalization is supported by embryology, by the fast-acting neurotransmitters used by projections from the cortex and nuclei, and (as we shall see in the next section) by the organization of descending projections from cerebrum to the motor system. What are the components of the cerebral nuclei, and how are they distributed between the striatum and pallidum? In addition to the topological features dictated by embryology (Figs. 8.1 and 8.9), it now seems clear that pyramidal cells, which generate the projections from the cortex, use the excitatory neurotransmitter glutamate, whereas the descending projections of the cerebral nuclei use the inhibitory neurotransmitter GABA. Thus, if it is unclear whether a particular cell group in the cerebral hemisphere is part of the cortex or the nuclei, the major neurotransmitter used in its descending projections is one criterion for helping to decide. The cerebral cortex also contains a large population of GABAergic neurons, but they are local circuit rather than projection neurons. John Rubenstein and his colleagues have recently demonstrated that most, if not all, of them are actually born in the cerebral nuclei region of the early embryo and then migrate dorsally into the cerebral cortex region. Thus, most, if not all, cerebral GABAergic neurons may be generated in the embryonic nuclear or basal region.

Based on these embryological and neurotransmitter utilization criteria, we can assign all noncortical cerebral cell groups to the cerebral nuclei; like the cerebral cortex, they are arranged in a topographically ordered way (Fig. 8.11). Further assignment to either the striatal or pallidal division of the cerebral nuclei may be done tentatively on the basis of known embryological relationships and, by analogy, to the connectional model of the classic striatum (caudate nucleus and putamen) and pallidum (globus pallidus), which we now review.

TRIPLE DESCENDING PROJECTION FROM CEREBRUM

Is there a basic minimal circuit that can be applied to all or most parts of the cerebral hemispheres? One appealing possibility involves its descending projection to the motor system (Fig. 8.12). It did not became clear until the 1960s that most of the isocortex generates a topographically organized projection to the entire striatum (the caudate and putamen, caudoputamen, or dorsal striatum). It is now known that this projection arises predominantly from layer 5 py-

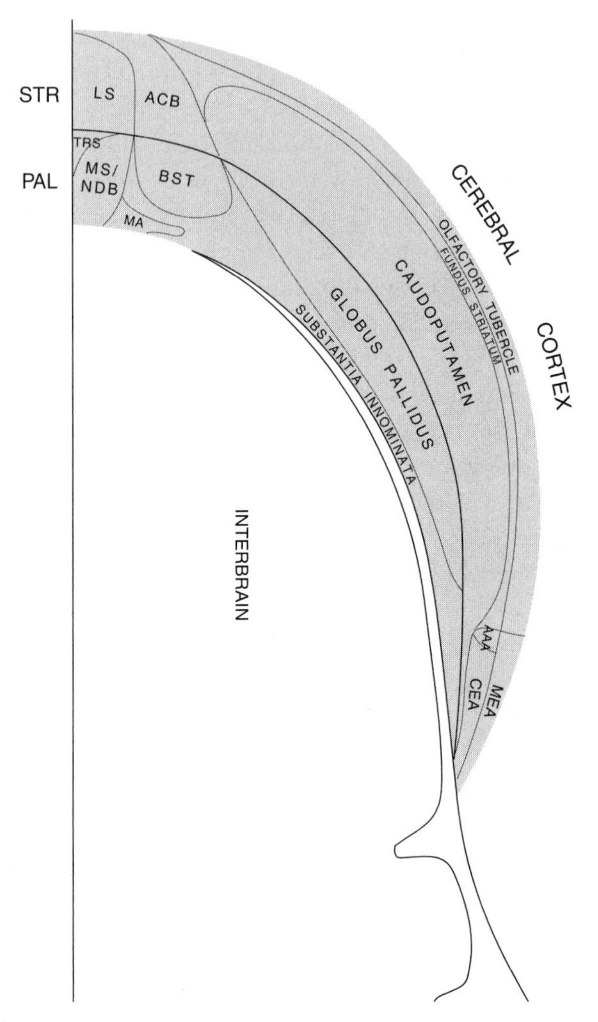

FIGURE 8.11 *The overall arrangement of cell groups within the cerebral nuclei (telencephalic basal ganglia) is indicated on this flatmap of the rat brain. Note that the cerebral nuclei may be divided into striatal (STR) and pallidal (PAL) domains.* Key: *AAA, anterior amygdalar area; ACB, nucleus accumbens; BST, bed nuclei of the stria terminalis; CEA, central amygdalar nucleus; LS, lateral septal complex; MA, magnocellular (preoptic) nucleus; MEA, medial amygdalar nucleus; MS/NDB, medial septal-nucleus of the diagonal band complex; TRS, triangular septal nucleus. Adapted from L.W. Swanson,* Brain Maps: Structure of the Rat Brain, *second edition (Elsevier Science: Amsterdam, 1998–1999).*

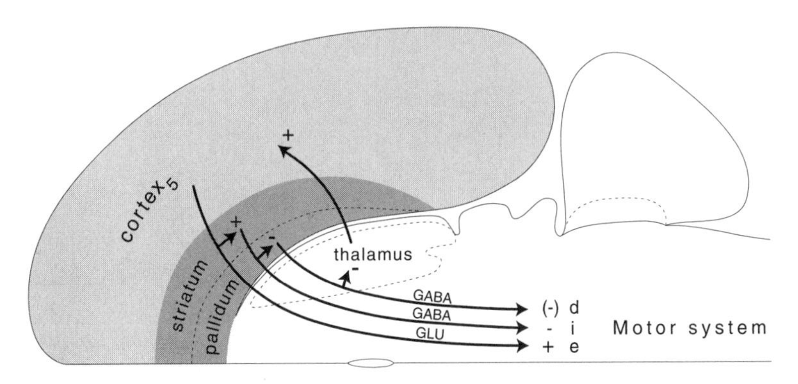

FIGURE 8.12 *The organization of a triple-descending projection from the cerebral hemispheres to the motor system, broadly defined. Note the thalamocortical feedback loop. In the isocortex, most descending projections arise from layer 5 pyramidal neurons.* Key: +, *excitatory (e); −, inhibitory (i); (−), disinhibitory (d). Adapted with permission of Elsevier Science from L.W. Swanson, Cerebral hemisphere regulation of motivated behavior,* Brain Res., *2000, vol. 886, p. 129.*

ramidal neurons that use the typically excitatory neurotransmitter glutamate. It is also known that this input is generated by collaterals of parent axons that descend at least as far as the brainstem. Thus, certain layer 5 cortical neurons provide simultaneous excitatory inputs to the caudate and putamen, as well as to the motor system broadly defined (Chapter 6).

The second part of the minimal cerebral circuit to the motor system involves the descending projections of the striatum (caudate and putamen), which use the typically inhibitory neurotransmitter GABA. This projection is relatively quite simple. The caudate and putamen together provide a topographically organized projection to the entire globus pallidus (both external and internal parts) and to both parts of the substantia nigra (reticular and compact). It has now been established by André Parent and others that the striatal projection to the globus pallidus arises as collaterals of parent axons that go on to the substantia nigra. Thus, individual neurons in the caudate and putamen provide a simultaneous inhibitory input to the globus pallidus and substantia nigra. Recall that the reticular substantia nigra is part of the behavior control column, concerned in part with generating orienting movements of the eyes and head (Chapter 6, Fig. 6.11).

The third and final part of the minimal cerebral circuit to the motor system involves the descending projections of the globus pallidus. As with the caudate and putamen, these projections use GABA as a neurotransmitter and generate a branched projection—this time to the motor system broadly defined, as well as to the dorsal thalamus. Because the globus pallidus receives an inhibitory input from the striatum, the descending pallidal projection to the motor system can be thought of as *disinhibitory* (an inhibitory projection is inhibited), a term introduced by Eugene Roberts, who discovered GABA in the nervous system. For example, both the striatum and the globus pallidus project heavily to the reticular part of the substantia nigra, as do certain regions of the cerebral cortex (motor-related areas of the frontal region). It seems likely that the direct cortical input is excitatory, the striatal input is inhibitory, and the pallidal input is disinhibitory (Fig. 8.12).

The minimal cerebral circuit under discussion so far has two essential components: a triple cascading projection to the motor system (broadly defined) and a thalamocortical feedback loop. Thus, cortical neural activity, which in some sense is equivalent to cognition, has a direct influence on the motor system, and the results of this influence are fed back onto the cortex to influence subsequent neural activity there. The next critical question in the analysis is this: Does this minimal cerebral circuit apply just to isocortex, caudoputamen, and globus pallidus—or is it characteristic of most or all of the cerebral hemisphere? My analysis of the connectional data now at hand, published in 2000, suggests that the latter is the case.

The first real breakthrough in this line of thinking was provided in 1975 by L. Heimer and R.D. Wilson. With brilliant insight, they suggested that there is a ventral striatum and ventral pallidum, which would complement the "classical" dorsal striatum (caudate and putamen) and dorsal pallidum (globus pallidus). The ventral striatum consisted primarily of the nucleus accumbens and olfactory tubercle, which project to the ventral pallidum (in the substantia innominata), and the latter, in turn, projects to the thalamus. Thus both the dorsal and ventral striatopallidum form a thalamocortical loop, although the projections are topographically organized so that different parts of the thalamus and cortex are influenced by the dorsal and ventral cerebral nuclei.

We have now extended this view to include all of the cerebral nuclei and virtually the entire cerebral cortical mantle as well. The suggested arrangement of cerebral nuclei is outlined in Figure 8.13. There appear to be two new features here: the medial and caudorostral divisions of the striatopallidum (or cerebral nuclei). In this scheme, the lateral septal complex is striatum for the hippocampal cortex (or Ammon's horn), a suggestion first made by Cajal a century ago, and the medial septal–diagonal band complex is the pallidum associated with the hippocampus and lateral septum. Furthermore, certain regions of the traditional amygdala that have GABAergic projection neurons—in particular, the central and medial nuclei—are regarded as striatum for certain regions of olfactory and visceral cortex, and the bed nuclei of the stria terminalis are re-

Cerebral nuclei	***Dorsal***	***Ventral***	***Medial***	***Caudorostral***
STRIATUM	CP	ACB FS OT	LSC	MEA CEA AAA IA
PALLIDUM	GPe GPi	SI MA	MS/NDB	BST

FIGURE 8.13 *Four topographic regions of the cerebral nuclei (striatum and pallidum) are shown in this figure.* Key: *AAA, anterior amygdalar area; ACB, nucleus accumbens; BST, bed nuclei of the stria terminalis; CEA, central amygdalar nucleus; CP, caudoputamen; FS, striatal fundus; GPe, external globus pallidus; GPi, internal globus pallidus; IA, intercalated amygdalar nuclei; LSC, lateral septal complex; MA, magnocellular preoptic nucleus; MEA, medial amygdalar nucleus; MS/NDB, medial septal/nucleus of the diagonal band complex; OT, olfactory tubercle; SI, substantia innominata. Adapted with permission of Elsevier Science from L.W. Swanson, Cerebral hemisphere regulation of motivated behavior,* Brain Res., *2000, vol. 886, p. 113.*

garded as the corresponding pallidum. These arrangements fit the overall scheme because both supposed pallidal regions (medial septum–diagonal band complex and bed nuclei of the stria terminalis) establish thalamocortical feedback loops, and both are derived from the embryonic medial ventricular ridge.

From the standpoints of embryology and adult connections, the cerebral hemispheres appear to form an integrated unit—which from the functional perspective is responsible for elaborating cognition and for transmitting cognitive influences to the motor, sensory, and behavioral state systems. The key to understanding the cerebral hemispheres lies in the arrangement of the structure-function regionalization map of the cerebral cortex. The thalamus projects in a topographically organized way to virtually all of the cerebral cortex, and, in turn, almost all of it projects in a topographic way to the cerebral nuclei. Superficial (supragranular) pyramidal cells establish an immensely complex network of connections between the various cortical areas in the hemispheres on both sides of the brain, and this network must play a critical role in elaborating various aspects of cognition. Deep (infragranular) pyramidal neurons receive inputs from superficial pyramidal neurons and send massive descending projections to the cerebral nuclei, brainstem, and spinal cord. That is, deep pyramidal neurons play a key role in transmitting the results of neural activity in the intracortical network to the motor, sensory, and behavioral state systems. We shall now turn to a consideration of how information about the external and internal environments enters the nervous system to influence the cognitive, motor, and behavioral state systems.

READINGS FOR CHAPTER 8

Brodmann, K. *Vergleichende Localisationslehre der Grosshirnrinde in ihren Prinzipien dargestellt auf Grund des Zellenbaues*. Barth: Leipzig, 1909. For English translation, see L.D. Garey, *Brodmann's "Localization in the Cerebral Cortex"* (Smith-Gordon: London, 1994). This should be required reading for anyone interested in the cerebral cortex.

Clarke, E., and Dewhurst, K. *An Illustrated History of Brain Function: Imaging the Brain from Antiquity to the Present*, second edition. Norman: San Francisco, 1996.

DeFelipe, J., and Jones, E.G. *Cajal on the Cerebral Cortex: An Annotated Translation of the Complete Writings.* Oxford University Press: New York, 1988.

Heimer, L., and Wilson, R.D. The subcortical projections of allocortex: similarities in the neuronal associations of the hippocampus, the piriform cortex and the neocortex. In: M. Santini (ed.), *Golgi Centennial Symposium Proceedings.* Raven Press: New York, 1975, pp. 173–193.

Jones, E.G., and Powell, T.P.S. An anatomical study of converging sensory pathways within the cerebral cortex of the monkey. *J. Anat.* 93:793–820, 1970.

Marin, O., and Rubenstein, J.L. A long, remarkable journey: tangential migration in the telencephalon. *Nat. Rev. Neurosci.* 2:780–790, 2001.

Meyer, A. *Historical Aspects of Cerebral Anatomy.* Oxford University Press: Oxford, 1971. This is an invaluable, scholarly account of the topic.

Peters, A., and Jones, E.G. (eds.) *Cerebral Cortex.* Plenum Press: New York, 1984–1999. This is a fourteen-volume collection of articles by a wide range of experts.

Swanson, L.W. Cerebral hemisphere regulation of motivated behavior. *Brain Res.* 886:113–164, 2000.

Williams, P.L. (Ed.) *Gray's Anatomy*, thirty-eighth (British) edition. Churchill Livingstone: Edinburgh, 1995. There is an excellent review here of cerebral hemisphere functional anatomy.

9

The Sensory System

Inputs from the Environment and the Body

Impressions conveyed by the sensitive nerves to the central organs are either reflected by them upon the origin of the motor nerves, without giving rise to true sensations, or are conducted to the sensorium commune, the seat of consciousness. . . . It is probable that there is in the brain a certain part or element appropriated to the affections, and the excitement of which causes every idea to acquire the intensity of emotion, and which, when very active, gives the simplest thought, even in dreams, the character of passion; but the existence of such a part or element cannot be strictly proved, nor its locality demonstrated.

—JOHANNES MÜLLER (1843)

In many ways the sensory system is the easiest for us to understand at an almost intuitive level. The eye has a lens like a camera and a retina for capturing focused visual scenes almost like film, and then those scenes are transmitted by the optic nerve to the brain where they are somehow converted into sensations and perceptions. This same principle basically applies for sounds detected by the ear, odors detected by the nose, tastes detected by the tongue, hunger pangs detected by the stomach, and tickles and pinches detected by the skin.

Unavoidably, various aspects of the sensory system have already been dealt with in earlier chapters. We have seen that during the course of evolution various ectodermal cells became specialized to detect a wide range of stimuli from the external and internal environments. These receptor cells have been called *exteroceptors* and *interoceptors*, respectively. Examples of stimuli include chemicals (and the corresponding receptor cells, called *chemoreceptors*), temperature (*thermoreceptors*), mechanical deformation (*mechanoreceptors*), light (*photoreceptors*), and osmolality (*osmoreceptors*). From a strictly introspective point of view, you are conscious of the fact that stimuli detected by the eye and ear produce two entirely different sensory modalities. Johannes Müller, quoted at the beginning of the chapter, ascribed this qualitative difference between the classic sensory modalities (touch, taste, smell, vision, and hearing) to "specific nerve energies": because of these energies, he said, stimulation of a particular sense organ generates its own particular sensation and no other. Today, most neuroscientists have an alternative explanation. Each sensory system reaches a different region of cerebral cortex, where the qualitatively different conscious experiences associated with each sensory modality are elaborated (see Fig. 8.7).

We have also seen earlier in this book that the sensory system projects to the motor and behavioral state systems, in addition to the cognitive system, where conscious awareness is elaborated (Chapter 5). Direct sensory inputs to the motor system produce reflex behaviors without conscious awareness. In this chapter we look more carefully at certain general features that characterize the sensory system, as well as at special features that distinguish between the various sensory modalities in mammals. No attempt is made here to describe in detail the fascinating architecture of the individual sensory organs or the detailed circuit organization of each of the sensory subsystems. The latter topics are covered nicely in any introductory neuroscience textbook. Unraveling the anatomy, physiology, and chemistry of the sensory systems was a crowning achievement of post–World War II neuroscience.

EVOLUTION AND DEVELOPMENT OF SENSORY NEURONS

We noted in Chapter 2 that bipolar sensory neurons probably first evolved in the outer body wall layer (the ectoderm, facing the external environment) of Cnidarians such as hydra. In more complex invertebrate animals like worms that have a central nervous system, the axon from sensory neurons in the ectoderm extends into a ganglion of the ventral nerve cord. There the parent axon typically bifurcates, sending one bifurcation branch rostrally and the other caudally—with each branch, in turn, generating relatively short collaterals that ramify in the nearby neuropil (see Figs. 3.2, 3.4, and 9.1A). In more advanced invertebrates, such as mollusks for example (Fig. 9.1B), the cell body of many sensory neurons has migrated during development into the interior of the animal, just deep to the ectodermal epidermis. This arrangement provides a competitive advantage because the trophic center (the nucleus with its chromosomes, and most of the protein synthetic machinery) of the sensory neuron is better protected from potential damage inflicted by insults from the external environment. Note that here the bipolar sensory neuron's receptive pole or process (the dendrite) becomes elongated.

Turning to vertebrates, we find that, typically, sensory neurons innervating the skin have a cell body that is located in a dorsal root ganglion deep within the body. In fact, dorsal root ganglia are very protected from injury—so much so that they lie within a pocket formed by the vertebrae themselves. However, it is very interesting to see that in the adult, these ganglion cells do not have a bipolar shape; instead, they have a rounded cell body with a single process extending out to form a T-shaped arrangement (Figs. 5.4 and 9.1C). This single process is the dendrite, and it extends peripherally through a mixed nerve toward the periphery (or the viscera). The central, thinner part of the T-shaped arrangement is the axon, which extends through a dorsal root into the spinal cord or through certain cranial nerves to the brainstem. These dorsal root ganglion neurons have what is called a *pseudounipolar* shape, because, as we shall now see, they develop from a typical bipolar shape in the embryo.

each cell begins to migrate at right angles to the two processes, toward the periphery of the ganglion (Fig. 9.2C, then B). This migration of the nucleus, along with the accompanying perikaryon, produces the definitive, adult pseudounipolar shape of the dorsal root ganglion neurons. The stem of the T is thick and extends uninterrupted to the periphery or viscera. Functionally it is the dendrite because it detects stimuli and conducts them toward the axon, which arises from the sharp bend of the dendrite within the ganglion. The axon is considerably thinner than the dendrite and conducts information into the central nervous system. In vertebrates, it is quite common for the axon of a neuron to arise from a dendrite. For example, this arrangement is found in cell types as varied as the cerebral cortical pyramidal neurons and the cerebellar cortical granule cells.

For the sake of completeness, we should recall (Chapter 4) that not all "dorsal root ganglion cells" are found in dorsal root ganglia. In 1877 Sigmund Freud discovered that in lampreys (a primitive vertebrate) some of these sensory neurons are found in ganglia near the spinal cord, and some are actually found within the spinal cord itself. Then a few years later Gustaf Retzius discovered that in a protovertebrate, amphioxus, most of the sensory neurons are found within the spinal cord. All of these sensory neurons are derived from the embryonic ectoderm, of course, but in the more advanced vertebrates the neural crest and epibranchial placodes are more clearly differentiated from the neural plate that goes on to generate the neural tube and thus the spinal cord.

OVERVIEW OF SENSORY NEURONS

Most sensory neurons in humans and other mammals are variations on the bipolar and pseudounipolar ganglion cells just discussed (see Fig. 9.3). Olfactory neurons are the closest to the primitive bipolar sensory neurons found in earthworms, for example (Figs. 3.4 and 9.1A). The cell bodies of olfactory neurons lie in the olfactory mucosa of the nose (Fig. 9.3A, B)—that is, in a specialized epithelial cell layer derived from the ectoderm. Their axon extends all the way to the brain, where it synapses with the dendrites of mitral cells in the olfactory bulb of the cerebral cortex (to be discussed later in this chapter).

In contrast, bipolar sensory neurons whose cell body has migrated deeper into the body as in mollusks (Fig. 9.1B) are represented in the auditory and vestibular systems, which use the eighth cranial nerve (Fig. 9.3D). Bipolar neurons whose dendrites extend to the organ of

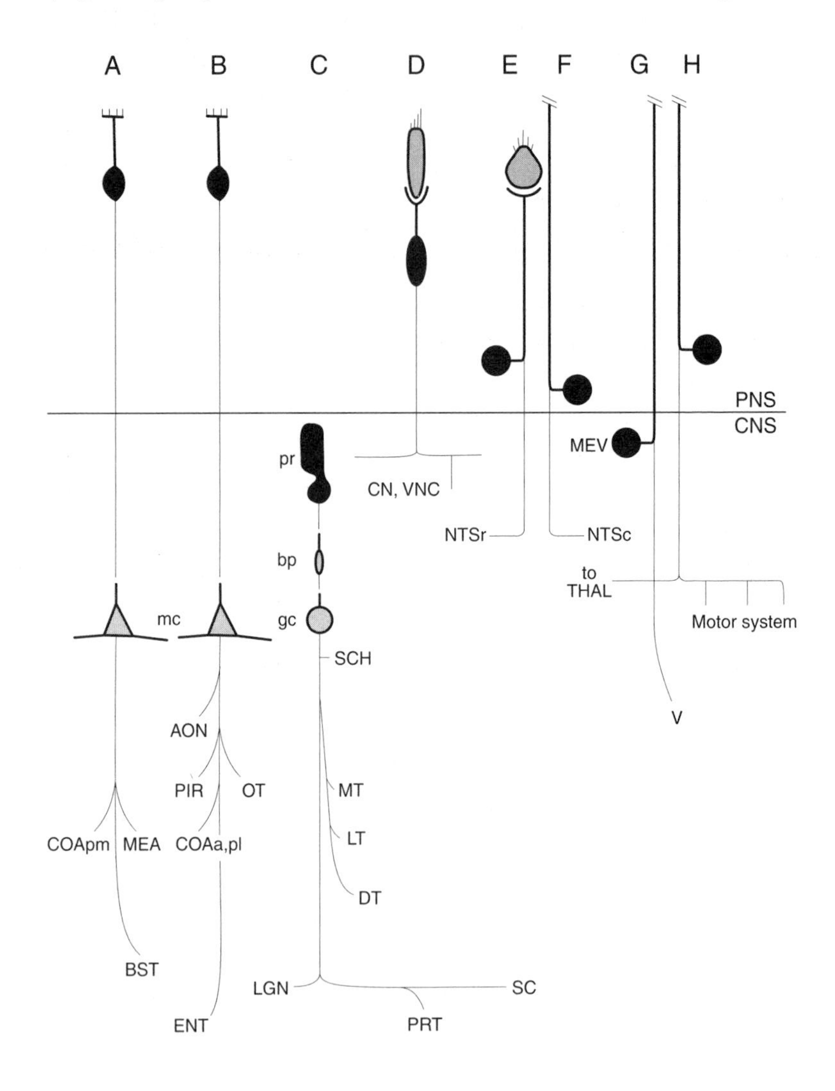

Corti in the cochlea are concentrated in the spiral ganglion, and their axons end in the cochlear nuclei of the brainstem. Bipolar neurons whose dendrites extend to the semicircular canals (the other half of the inner ear) are concentrated in the vestibular ganglion, and their axons end in the vestibular nuclei of the brainstem. Interestingly, the dendrites of eighth nerve bipolar neurons are myelinated and conduct action potentials (nerve impulses).

Stimuli associated with hearing and equilibrium are detected by sensory cells called *hair cells* in the organ of Corti and semicircular canals, respectively. Hair cells are mechanoreceptors whose "hairs" (actually microvilli and cilia) detect pressure changes in the fluid surrounding them. Although they are sensory cells, hair cells are not traditionally regarded as neurons. They are derived from non-neural epithelium that lines the cavity of the inner ear (a lining called the *membranous labyrinth*). Taste cells are another example of non-neural sensory cells that are generated from a specialized epithelium, in this case associated with the tongue and nearby tissues.

Pseudounipolar sensory neurons are very common in mammals (Figs. 9.1C and 9.3E–H). The prototype (Fig. 9.3H) is the dorsal root ganglion cell, which is associated with each spinal nerve (see Figs.

FIGURE 9.3 *This schematic diagram shows the basic arrangement of the various sensory neurons (black) found in mammals.* Key: *A, vomeronasal neurons; B, olfactory mucosa neurons; C, photoreceptors; D, spiral and vestibular ganglion cells receiving information from hair cells; E, geniculate, glossopharyngeal, and vagal ganglion cells receiving information from taste receptors; F, glossopharyngeal and vagal ganglion cells innervating the viscera; G, midbrain nucleus of the trigeminal nerve; H, dorsal root ganglion cells. AON, anterior olfactory nucleus; bp, retinal bipolar cell; BST, bed nuclei stria terminalis; CN, cochlear nuclei; CNS, central nervous system; COAa, pl, pm, cortical amygdalar nucleus, anterior, posterolateral, posteromedial parts; DT, dorsal terminal nucleus, accessory optic tract; ENT, entorhinal area; gc, retinal ganglion cell; LGN, lateral geniculate nucleus; LT, lateral terminal nucleus; mc, mitral cell; MEA, medial amygdalar nucleus; MEV, midbrain nucleus of the trigeminal nerve; MT, medial terminal nucleus; NTSc,r, nucleus of the solitary tract, caudal, rostral parts; OT, olfactory tubercle; PIR, piriform area; PNS, peripheral nervous system; pr, photoreceptor; PRT, pretectal region; SC, superior colliculus; SCH, suprachiasmatic nucleus; THAL, thalamus; V, trigeminal motor nucleus; VNC, vestibular nuclei.*

5.4 and 6.4), along with the fifth cranial nerve trigeminal ganglion cell, which is analogous to a dorsal root ganglion cell for the skin of the head. The fact that the axon of dorsal root ganglion neurons bifurcates into ascending and descending branches on entering the spinal cord was discovered and described in 1885 by the great Norwegian scientist, arctic explorer, politician, and philanthropist Fridtjof Nansen, who won the Nobel Prize in 1922. It was part of his thesis work on the lowly hagfish (*Myxine glutinosa*) at the Bergen Museum. Soon thereafter, Cajal went on to show that this bifurcation is also universally true in birds and mammals and that, in turn, the bifurcation branches generate abundant collaterals that end in the spinal gray matter. Thus, the terminals of a dorsal root ganglion cell can contact a large number of postsynaptic neurons, including motoneurons, interneurons associated with motor pattern generators, and neurons with ascending projections to various parts of the brainstem, including the thalamo-cortical projection system (see Fig. 5.4). There is very extensive divergence of information transmission at the first stage of the dorsal root or trigeminal ganglion sensory system.

There are other variations on the pseudounipolar sensory neuron theme. One example involves neurons in the ganglia of the ninth (glossopharyngeal) and tenth (vagus) cranial nerves (Fig. 9.3F). These neurons detect a wide range of sensory information from the viscera (and very limited amounts from the soma) and transmit this information to the caudal end of a hindbrain sensory nucleus, the nucleus of the solitary tract (for somatic information, the adjacent trigeminal nucleus). Unlike the axon of dorsal root ganglion cells, the axon of these vagal and glossopharyngeal ganglion cells typically does not bifurcate. Instead, it enters the brainstem and extends caudally through the solitary tract, where it branches to innervate the nucleus associated with the tract. Another example involves the pseudounipolar sensory neurons that innervate taste buds (Fig. 9.3E). These neurons are found in three ganglia: the distal ganglia of the ninth and tenth cranial nerves, and the geniculate ganglion of the intermediate part of the seventh cranial nerve. Their axon courses rostrally in the solitary tract to innervate the rostral end of the tract's nucleus, which is specialized for gustation rather than viscerosensation. The final example involves the midbrain nucleus of the trigeminal nerve, which is a "dorsal root ganglion" in the brain, on the edge of the midbrain periaqueductal gray matter (Fig. 9.3G). It forms the afferent side of stretch reflexes that help control the muscles of mastication or chewing, which are innervated by the motor nucleus of the trigeminal (fifth cranial) nerve.

The last type of sensory neuron we come to is the photoreceptor of the eye. Curiously, perhaps, the retina, which contains a monolayer sheet of photoreceptors, is an outgrowth of the brain—of the hypothalamus in fact (see Chapter 4). Thus photoreceptors are central neurons. Their axon synapses with or innervates the dendrites of retinal bipolar cells (local circuit interneurons), whose axon, in turn, synapses with retinal ganglion cells, which are the projection neurons that form the optic nerves and tracts transmitting the results of retinal visual information processing to the rest of the brain. Photoreceptors are incredibly sensitive. Apparently they can detect and thus respond to a single photon.

OVERVIEW OF SENSORY PATHWAYS

The basic plan of the nervous system outlined in this book stresses the fact that the sensory system as a whole projects to the motor system, to the behavioral state system, and to the cognitive or cerebral system (see Fig. 5.5). Exactly how this is accomplished for each particular sensory modality is beyond our scope here. Instead, in the following sections we outline the major features of each modality. From this perspective it appears safe to conclude that there are few if any generalizations that apply to all of the various sensory modalities. In fact, attempts to formulate such generalizations or principles in textbooks have led to needless confusion and misconceptions. For example, not all sensory information reaches the cerebral cortex by way of a "relay" in the thalamus. Instead, olfactory sensory neurons project directly to the cerebral cortex (the olfactory bulb)—and visceral sensory information from the nucleus of the solitary tract can reach the cortex directly, as well as by way of a relay in the thalamus. Information from each sensory modality reaches the cerebral cortex in a different way (Fig. 9.4).

We see in Chapter 6 that the motor system core (excluding the cerebellum) may be analyzed effectively in terms of a hierarchical organization scheme. The sensory system is different. For a considerable distance, each of the pathways associated with various sensory

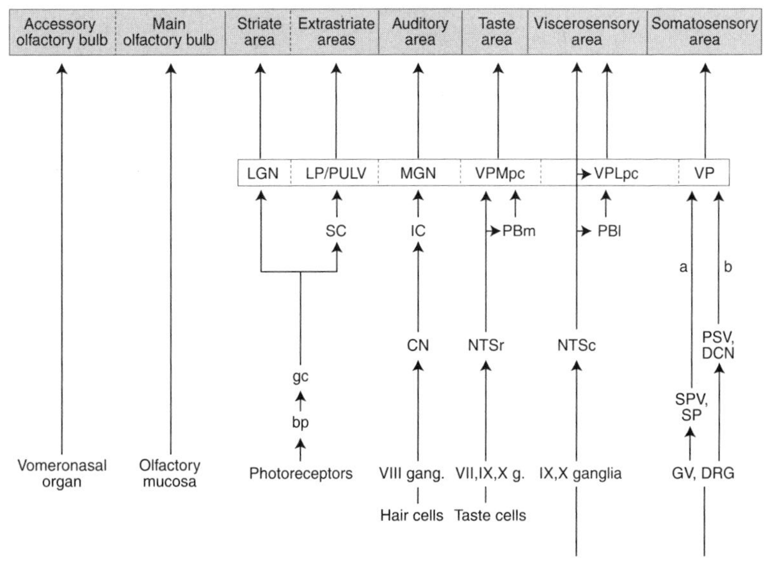

FIGURE 9.4 *Routes for different classes of sensory information to reach the cerebral cortex (shaded rectangle) are illustrated schematically in this diagram (unshaded rectangle is thalamus).* Key: *a, b, spinothalamic tract and medial lemniscus, respectively; bp, retinal bipolar cell; CN, cochlear nuclei; DCN, dorsal column nuclei; DRG, dorsal root ganglion; IX,X ganglia, glossopharyngeal and vagal ganglia; gc, retinal ganglion cell; IC, inferior colliculus; GV, trigeminal ganglion; IX, X ganglia, glossopharyngeal and vagal ganglia; LGN, lateral geniculate nucleus; LP/PULV, lateral posterior nucleus/pulvinar; MGN, medial geniculate nucleus; NTSc, r, nucleus of the solitary tract, caudal, rostral parts; PBl, m, parabrachial nucleus, lateral, medial divisions; PSV, principal trigeminal sensory nucleus; SC, superior colliculus; SP, spinal cord; SPV, spinal trigeminal nucleus; VII, IX, Xg., gerriculate, petrosal, and nodose ganglia; VIII gang., vestibulocochlear ganglion; VP, ventral posterior nucleus; VPLpc, ventral posterolateral nucleus, parvicellular part; VPMpc, ventral posteromedial nucleus, parvicellular part.*

modality classes remains separate. That is, they are arranged in parallel, at least in terms of the primary modality classes (vision, audition, olfaction, taste, and somatovisceral). Convergence takes place in polymodal regions of the cerebral hemispheres and in specialized regions of the brainstem—most (if not all) of which are parts of the cognitive, behavioral state, and motor systems.

FOREBRAIN SENSORY SYSTEMS
Olfactory, Visual, Humoral, and Osmotic

The forebrain is the most complex part of the nervous system, and it has a number of special sensory systems that are unique and have an important role in generating this level of complexity. The extent to which these sensory systems are differentiated relative to one another (their relative size in crude terms) varies greatly between various species. For example, the olfactory sense is very prominent in mice and relatively paltry in humans, whereas the reverse is true for the visual system: humans have a relatively elaborate visual system, as compared to nocturnal, almost blind mice that tend to avoid light whenever possible. The olfactory and optic nerves are the major cranial nerves of the forebrain. Their relative sizes vary greatly between humans and mice, and yet both sensory systems have the same basic organization, the same fundamental plan, not only in both species but in all mammals generally. This is simply a variation on a theme emphasized in each chapter: there is a fundamental plan of the mammalian nervous system, and the nervous system in each species is a quantitative variation on that theme.

Let us begin with the first cranial nerve, the olfactory nerve. As we have already seen, this nerve has very primitive—or perhaps more accurately, very ancient and thus very conserved—features and is in many ways the simplest of the sensory systems in terms of network organization (see Fig. 9.4). However, before we get to the main olfactory system, it is important to realize that the traditional classification of 12 cranial nerves (Chapter 6) breaks down completely here because there are at least three nerves associated with the rostral end of the central nervous system.

Although the olfactory nerve is cranial nerve I, there is also a distinct vomeronasal nerve that is unnumbered. It arises from a pit in the olfactory mucosa where the bipolar sensory neurons are specialized to detect pheromones—molecules released into the air by one animal and detected by another animal, to influence various aspects of social behavior (for example, sexual, parental, and territorial behaviors). The vomeronasal nerve provides a beautiful example of major differences between species. The vomeronasal system is very prominent in rodents, where it is essential for reproductive behaviors, but it is absent in the vast majority of adult humans. However, this absence in the adult human is the result of atrophy; it does not reflect a fundamental difference in terms of a complete absence in one species. The vomeronasal system develops in the human embryo just as it does in rodents, but later on it degenerates (for unknown reasons).

In addition, there is an enigmatic terminal nerve that was discovered in the second half of the nineteenth century, but it also does not have a number in the 12 cranial nerve scheme. It appears to innervate the nasal mucosa and send fibers into the brain in the region of the terminal lamina; that is, in the medial septal–diagonal band complex of the cerebral nuclei and the adjacent preoptic region of the hypothalamus. Recently, the terminal nerve has attracted attention because it probably forms the route taken by GnRH neuroendocrine neurons migrating from their embryonic birthplace in the olfactory epithelium to their final resting place in the basal forebrain (Chapter 6).

The initial stages of the vomeronasal system are very simple indeed. The vomeronasal nerve ends in the accessory olfactory bulb, a specialization of the main olfactory bulb where the olfactory nerve ends in the brain. Topologically and embryologically, the main and accessory olfactory bulbs are the earliest differentiations of the cerebral cortex (see Fig. 8.5), and various authorities—including Cajal and Brodmann—have thought with good reason that these two parts of the bulb are primary sensory cortices for the main and accessory olfactory systems, respectively. In other words, the vomeronasal nerve ends entirely and thus exclusively in the primary vomeronasal cortex: the accessory olfactory bulb (Figs. 9.3 and 9.4).

Fortunately, the projections of the primary vomeronasal cortical area are very simple and follow the most basic projection pattern of other unimodal sensory areas (see Fig. 8.12). The accessory olfactory bulb projects massively to an association vomeronasal cortical area (the posteromedial cortical nucleus of the amygdala) and a region of the striatum (the medial amygdalar nucleus; Fig. 8.11). To complete the picture, the accessory bulb projects lightly to a region of the pallidum (the principal nucleus of the bed nuclei of the stria terminalis or BST, which receives a massive input from the striatal component, the medial amygdalar nucleus). Thus, the accessory olfactory system participates in a classic triple-descending projection to the motor system (Fig. 8.12) from the cerebral cortex, the striatum, and the pallidum. Its major inputs to the motor system involve the hypothalmus—in particular, the rostral group of medial nuclei that control the expression of social behaviors, and the visceromotor pattern generator network next to it in the periventricular region (see Fig. 6.11).

The main olfactory system is more complex because its primary sensory cortical area (the main olfactory bulb) has much more widespread projections to secondary olfactory cortical areas. Among others, they include the anterior olfactory nucleus, piriform area, anterior and posterolateral parts of the cortical amygdalar nucleus, and entorhinal area of the hippocampal formation (Fig. 9.3). The striatal projection of the primary olfactory cortex (main olfactory bulb) is primarily to the olfactory tubercle, which lies just rostral to the accessory olfactory striatum (the medial amygdalar nucleus), and the corresponding region of pallidum is centered in restricted (rostroventral) parts of the substantia innominata (ventral pallidum). The secondary main and accessory olfactory cortical areas are not strictly unimodal: they receive information from each other and thus, to some extent, at least, integrate information from both differentiations of the olfactory system. The primary main olfactory cortical area does not project to the brainstem; instead, the secondary main olfactory cortical areas (especially those in the amygdalar region) and the substantia innominata carry olfactory information to the brainstem, mainly to the hypothalamus, and restricted parts of the thalamus.

There is an enormous literature on the visual system, and we can only touch on selected highlights here. To start with, the processing of visual information in the retina is exceptionally complex. It is well known that bipolar neurons "relay" information from photoreceptors to retinal ganglion cells, which, in turn, send this information through the optic nerves and tracts to the rest of the brain (Fig. 9.3). In addition, there are two layers of other interneurons (horizontal and amacrine cells), and they spread information from photoreceptors tangentially through the retina. Thus, there are five basic neuronal cell types in the retina, and the actual details of information processing within the network that they form are vaguely understood.

One reason for this lack of understanding is the extensive differentiation of each of the five retinal neuronal cell types into subtypes or varieties with important functional differences. For example, there are two classes of photoreceptors, rods and cones, which are responsible for night and day (color) vision, respectively. In addition, there are three types of cones that are maximally sensitive to red, green, and blue wavelengths of light. Finally, there are at least three major types of bipolar cell, at least six major types of ganglion cell, and dozens of varieties of horizontal and amacrine cells.

The optic nerve terminates quite extensively in the brain. Its first offshoot is to the suprachiasmatic nucleus, which lies just dorsal to the optic chiasm (where the optic nerves from each eye cross or partly cross to the other side of the brain) and is the brain's primary circadian rhythm generator or clock (Chapter 7 and Fig. 9.3). Just beyond this level, an offshoot of the optic tract (the name for the continuation of the optic nerve beyond the chiasm), the accessory optic tract, splits off and courses to the midbrain, where it ends in three terminal nuclei (medial, lateral, and dorsal). These nuclei play an important role in controlling eye movements and are thus parts of the motor system. The main optic tract continues on to end in the superior colliculus or optic tectum of the midbrain, after giving off collaterals to the lateral geniculate nucleus of the thalamus and to the pretectal region (Fig. 9.3). The dorsal part of the lateral geniculate nucleus then projects to the primary visual cortex, whereas the pretectal region is involved in visual reflexes and the superior colliculus has two main roles: projecting to the motor system and projecting to secondary visual cortical areas via the thalamus (Fig. 9.4). Other less prominent, and less understood, terminal fields of the optic nerve in mammals include the lateral hypothalamic area, anterior thalamic nuclei, bed nuclei of the stria terminalis, and dorsal raphé nucleus.

The subfornical organ is an embryonic differentiation of the forebrain roof plate, in a dorsal region between the interbrain (thalamus) and endbrain. This nucleus lacks the normal blood–brain barrier so that its neurons are exposed directly to peptide hormones in the blood. One such hormone is angiotensin II, whose levels go up when there is a loss of body fluid because of dehydration or hemorrhage. Under these conditions, blood pressure needs to be maintained and water needs to be ingested. Neurons in the subfornical organ have angiotensin II receptors, and when they are activated three subfornical pathways to other parts of the brain are activated. One pathway modulates hypothalamic inputs to medullary autonomic baroreceptor reflex centers that control blood pressure; another pathway modulates the release of hypothalamic neuroendocrine hormones that regulate body water retention and blood pressure, and yet another pathway stimulates thirst and drinking behavior. Thus, the subfornical organ is a "humerosensory" organ or nucleus that detects hormone levels in the blood. Like the retina, its sensory neurons are derived from the brain.

Finally, we come to the osmoreceptors of the hypothalamus. It has been known since the classic studies of Earnest Basil Verney and Bengt Andersson starting in the 1940s that they are found in the rostral end of the hypothalamus, around the rostral end of the third ventricle, and that they are responsible for eliciting drinking and the secretion of hypothalamic neuroendocrine hormones that regulate body water. Their precise cellular identity remains a mystery, but it is clear that they respond to increased osmolality of the blood due to loss of body water. There is good reason to believe that the subfornical organ projects to these preoptic osmoreceptors, so that the subfornical organ angiotensin-sensing neurons and the preoptic osmosensitive neurons work together as part of a system to control drinking behavior and body water regulation.

GANGLION CELL SENSORY SYSTEMS *Submodalities*

The sensory ganglion cells of four sensory systems illustrated in Figure 9.3D–F send their axons to primary sensory nuclei in the dorsal medulla. We are referring here to the ganglion cells of the (a) auditory system, which end in the cochlear nuclei; (b) vestibular system, which end in the vestibular nuclei; (c) gustatory system, which end in the rostral nucleus of the solitary tract; and (d) vagal/glossopharyngeal visceroceptive system, which end in the caudal nucleus of the solitary tract. These special sensory nuclei are all derived in the embryo from a highly differentiated, dorsal region of the hindbrain vesicle, the rhombic lip (Chapter 4, Fig. 4.15). Without going into details, these sensory nuclei generate axonal projections or pathways to the cognitive/cerebral (Fig. 9.4), behavioral state, and motor systems.

Finally we come to the dorsal root ganglion system associated with the spinal nerves and their serial homolog in the cranial region, the trigeminal ganglion. The sensory nuclei of the trigeminal nerve develop just below (ventral to) the rhombic lip in the embryo, and for all intents and purposes these nuclei represent a rostral extension into the brainstem of the corresponding regions of the spinal cord that receive inputs from the dorsal root ganglia.

The dorsal root ganglion system is commonly equated with the somatic sensory system, but the meaning of the latter term needs to be clear. Dorsal root ganglia transmit sensory information from what is usually thought of as the soma, or body (by and large the skin and skeletomotor system), as well as from the viscera. In this, they transmit a rather diverse array of sensory modalities, including touch, pain, temperature (from hot to cold), muscle and tendon stretch, and the state of joints, ligaments, and fascia. Stretch receptors in muscles and tendons are unusual in that they participate in reflexes that control muscle tone, but their activity does not seem to reach the level of consciousness. In contrast, sensory information from the joints, ligaments, and fascia is important for elaborating the kinesthetic sense—the conscious awareness of body position in space. The splanchnic nerves (Chapter 6) contain dorsal root ganglion processes that transmit a wide range of sensations from the viscera to the spinal cord, and then the brain. The splanchnic nerves also contain abundant preganglionic axons to the paravertebral sympathetic ganglia (Fig. 6.12).

The peripheral ends of dorsal root ganglion cells display a wide range of appearances, from completely unelaborated and naked to having elegant encapsulations by surrounding non-neuronal cells (Fig. 9.5). In general, the simple peripheral fibers are associated with small dorsal root ganglion cells; they conduct action potentials slowly, detect thermal and painful stimuli, and tend to reach the thalamus via the spinothalamic tracts (Fig. 9.4a). In contrast, encapsulated peripheral endings are associated with large dorsal root ganglion cells; they conduct action potentials much faster, detect touch and stretch stimuli, and preferentially reach the thalamus via the dorsal column–medial lemniscal pathway (Fig. 9.4b).

In other words, the "somatic sensory system" or dorsal root ganglion system actually has a number of submodalities that have more or less distinct pathways within the spinal cord and to the cerebral cortex. Nevertheless, multimodal dorsal root ganglion cells do exist, and there is extensive convergence of somatic and visceral inputs onto individual neurons at all levels of the spinal cord and brain. The situation is not so different in principle from the other major sensory modalities. There are main and accessory olfactory systems; there are visual "subsystems" for the four types of photoreceptors; and there are separate nerves and ganglia for the two major divisions of the inner ear—the cochlea and semicircular canals. In the end, each sensory modality has its own differentiations, although all of the major

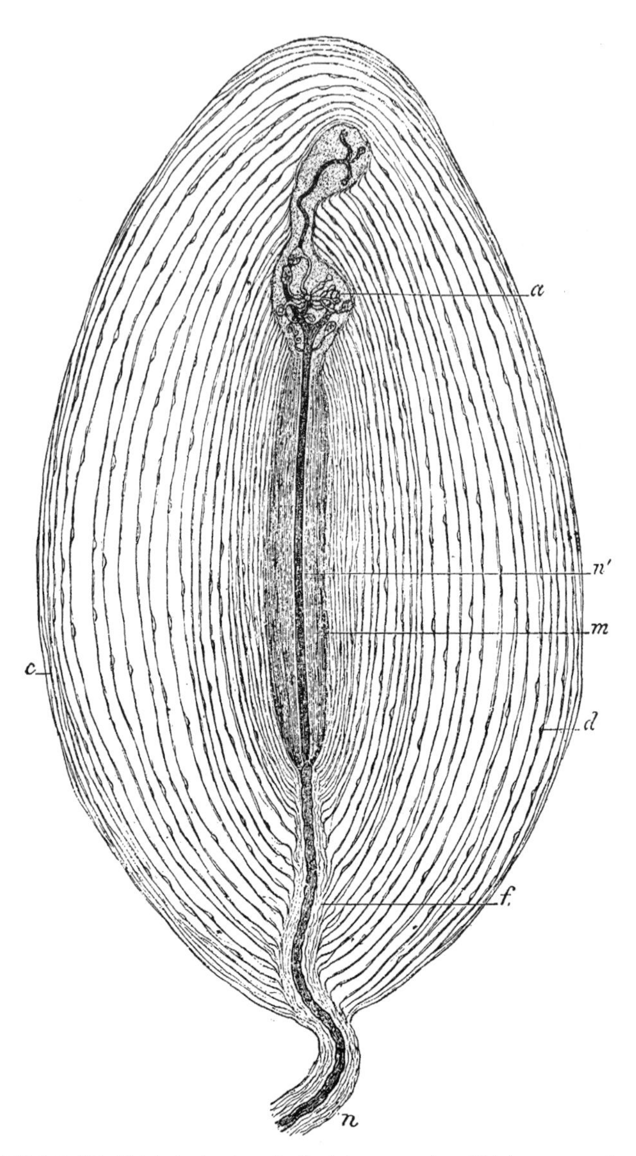

FIGURE 9.5 *The histologic structure of a Pacinian corpuscle, which is an encapsulated nerve ending that is a mechanoreceptor especially sensitive to vibration. The corpuscle is from the mesentery of a cat and is shown in cross-section; the lamellae are formed by specialized fibroblasts. The illustration is based on a drawing by Louis Ranvier, 1875.* Key: *a, region where one of the branches of the terminal fiber divides into many branches that go on to form abundant terminal boutons; f, perineural sheath; m, central mass or inner bulb; n, nerve fiber leaving the capsule; n′, terminal fiber. From L.F. Barker,* The Nervous System and Its Constituent Neurones *(Appleton: New York, 1901).*

types establish more or less parallel pathways to specific, primary sensory areas of the cerebral cortex (Fig. 9.4).

AFFECT *Pain and Pleasure, Emotion, and Mood*

Motivation and emotion are two fundamental aspects of conscious experience that are standard topics in psychology and are making a comeback in respectable neuroscientific circles. A vast array of experimental data provide clues to the localization of neural mechanisms that subserve motivation and emotion, and yet, in all candor, it must be admitted that the basic organization of these mechanisms remains elusive and unknown. We broached the topic of motivation or drive in Chapter 7, under the discussion of motor pattern initiators. There it was suggested that intrinsic neural activity in the hypothalamus plays an important role in motivation, although the crucial question of where the conscious perception of drive or motivation is elaborated remains unanswered. Is it cortical or subcortical?

Emotion leads us into the more general topic of affect, or feeling. Tremendous interest in the cognitive or thinking aspect of consciousness is evident today, although the affective aspect is really what separates us from machines like computers that are much better than we are at logical operations. A good case can be made for the proposition that all conscious experience is accompanied by affective tone, ranging from subtle feelings of comfort or discomfort to extreme emotions like rage and orgasm. What neural systems subserve affect, and where are the perceptions actually elaborated in the brain?

Perhaps a good place to start is with pain and pleasure. After all, they are the conscious reflections of punishment and reward, the reinforcement for learning whether to avoid or repeat a behavioral act. Behaviors associated with pain tend to be avoided in the future, and those associated with pleasure tend to be repeated. In addition, the organization of pain or nociceptive neural systems has been studied in great detail. As noted in the previous section, painful stimuli are detected by small sensory ganglion cells with thin, simple peripheral processes to the body wall and viscera. These stimuli are transmitted to the brain from spinal neurons that generate the so-called spinothalamic tracts. In fact, the spinothalamic tracts have extensive projections in the brain, with offshoots to the brainstem reticular formation and periaqueductal gray and to the nucleus of the solitary tract and parabrachial nuclei, before ending extensively in the thalamus. In addition, the tracts extend to innervate restricted parts of the hypothalamus, the cerebral nuclei, and even medial prefrontal regions of the cerebral cortex.

All of these projections of the spinothalamic tracts are important. But for now we should note that terminal fields in the thalamus include the ventral posterior nucleus, which also receives touch information through the dorsal column system and projects to the classic somatosensory cortex (Fig. 9.4); and the posterior complex, and midline and intralaminar nuclei—which together have very widespread cortical projections. Overall, it is clear that the central nociceptive system involves many structures with an accompanying circuitry that is very complex. And if the elaboration of consciousness is a property of the cerebrum, there are many candidate cortical areas because of the widespread distribution of nociceptive information in the thalamus. Fortunately for neuroscience at least, people with lesions in various cortical regions have provided important clues as to where the perception of pain may be elaborated. This evidence—as well as very recent data obtained with functional imaging in neurologically intact, unanesthetized people—would suggest that pain is perceived as such in a region of cortex that includes either the prefrontal region and/or the rostral half of the insular region, or both. People with lesions in this general region report that they can feel they are being pinched or poked, but that it does not hurt.

Relatively little work has been done on pleasure systems in the brain. Perhaps the most interesting aspect of this problem began with the discovery by James Olds in the 1950s that there are parts of the brain that rats will voluntarily stimulate electrically if an electrode is placed within them. Animals will self-administer this electrical stimulation of the brain thousands of times a day, to the exclusion of eating, drinking, and all other behaviors; and Robert Heath, a neurosurgeon, showed in the 1960s that people report pleasurable feelings akin to orgasm when electrodes are placed in certain forebrain regions. Apparently this experimental procedure is tapping into a pleasure system in the brain.

It now appears that the pleasure system accessed by the electrical self-stimulation paradigm involves pathways stretching between the prefrontal-insular cortex and ventral cerebral nuclei, through the lateral hypothalamic area, and into medial regions of the midbrain and pontine tegmentum (reticular formation). One reason for medical interest in this system is the anhedonia or loss of affect associated with clinical depression. Analogous to the clinical data just discussed for pain, it has been found that lesions in the medial prefrontal cortex can produce anhedonia in people, and very recent functional imaging studies also tend to confirm this localization.

Taken together, available evidence currently suggests that the perception of pain and pleasure is elaborated in a band of cerebral cortex that includes the medial prefrontal region and caudally adjacent parts of the insular region. This is especially interesting in view of the fact that visceral and nociceptive information from the nucleus of the solitary tract and parabrachial nucleus reaches this same band of cortex via both direct projections and projections relayed through the thalamus. Think about what an emotional experience is from an introspective point of view. Emotional experiences are "from the heart" or "from the gut." Visceral sensations are virtually synonymous with emotion. They involve changes in the perception of heart rate, breathing, and body temperature. The most obvious explanation is that they involve perception of visceral sensations.

Thus, the conscious perception of pain, pleasure, and emotions may be elaborated in a band of prefrontal-insular cortex. Based on known connections of this cortical band, it seems obvious that neural activity there can be influenced either by sensory inputs or by inputs from association cortical areas. Thus, affective experience may be evoked either by sensory inputs or by activity in association cortex (for example, during thinking or dreaming).

Moods are different. They have much longer durations that stretch into days and weeks, rather than being transitory affective responses to one type of stimulus or another. And mood tends to be stable, almost as if a set-point has been changed. There is, of course, no known explanation for the regulation of mood in terms of neural systems. However, it seems reasonable to suggest that activity in one or more of the behavioral state control nuclei that project to the prefrontal-insular region of the cerebral cortex (Chapter 7, Fig. 7.5) play a critical role. And recent experimental work indicates that an intracerebral neural network involving bidirectional connections between the prefrontal-insular cortex, the basal and central amygdalar nuclei, and the hippocampal formation is responsible at least in part for mediating learned emotional responses. From what has been said, it seems clear that the outlines of neural systems that elaborate affect are gradually coming into focus, but they are much more enigmatic than the classic sensory systems involving vision, audition, touch, and taste.

READINGS FOR CHAPTER 9

Bechara, A., Damasio, H., and Damasio, A.R. Emotion, decision making and the orbitofrontal cortex. *Cereb. Cortex* 10:295–307, 2000.

Berthier, M., Starkstein S., and Leiguarda, R. Asymbolia for pain: a sensory-limbic disconnection syndrome. *Ann. Neurol.* 24: 41–49, 1988.

Cajal, S.R. *Histologie du système nerveux de l'homme et des vertébrés*, 2 vols. Maloine: Paris, 1909–1911. For American translation, see *Histology of the Nervous System of Man and Vertebrates*, translated from the French by N. Swanson and L.W. Swanson, 2 vols. (Oxford University Press: New York, 1995). Cajal provides a nice overview to thinking about sensory system organization at the beginning of the twentieth century, especially from the perspective of cell morphology and embryology.

Finger, S. *Origins of Neuroscience: A History of Explorations into Brain Function.* Oxford University Press: New York, 1994. This is an excellent introduction to the historical development of ideas about the various sensory systems.

Gusnard, D.A., Akbudak, E., Shulman, G.L., and Raichle, M.E. Medial prefrontal cortex and self-referential mental activity: relation to a default mode of brain function. *Proc. Natl. Acad. Sci. USA.* 98:4259–4264, 2001.

Handbook of Sensory Physiology. Springer-Verlag: Berlin, 1971–. Nine authoritative volumes have been published so far; a terrific resource.

Petrovich, G.D., and Swanson, L.W. Combinatorial amygdalar inputs to hippocampal domains and hypothalamic behavior systems. *Brain Res. Rev.* 38:247–289, 2001.

Schafe, G.E., Nader, K., Blair, H.T., and LeDoux, J.E. Memory consolidation of Pavlovian fear conditioning: a cellular and molecular perspective. *Trends Neurosci.* 24:540–546, 2001.

Swanson, L.W. Cerebral hemisphere regulation of motivated behavior. *Brain Res.* 886:113–164, 2000.

Williams, P.L. (Ed.) *Gray's Anatomy*, thirty-eighth (British) edition. Churchill Livingstone: Edinburgh, 1995. Here one finds an authoritative review of the classic sensory systems.

10

Modifiability

Learning, Stress, Cycles, and Damage Repair

> Current methods and ideas are entirely dependent on continuing progress in chemistry and physics, which remain the principal allies of the naturalist.
>
> —SANTIAGO RAMÓN Y CAJAL (1911)

While it can be useful to think of the brain as a biological computer, it is a grave mistake to think about the brain's circuitry in terms of hard-wired computer chips. In the first place, because of the chromosome mixing that accompanies sexual reproduction, every brain is different in detail, just as every face is unique. And in the second, the brain is an organ composed of tissues; as such, it is alive and constantly changing. It was a major breakthrough in late-twentieth-century neuroscience to realize that the chemistry of synaptic transmission is dynamic, that new neurons may be generated in adult mammals, and that it may become possible to regrow damaged axonal pathways.

The topic of neural plasticity is vast and is only touched on here to emphasize, with selected examples, the principle that brain architecture is not static. The structure of the brain is constantly changing because of influences from the internal and external environments and because of genetic factors associated with the normal life cycle of development, puberty, adulthood, and aging. Nevertheless, it is important to realize that there are two qualitatively different levels of brain organization: macrocircuitry and microcircuitry.

The macrocircuitry of the brain can be thought of as the gross anatomy level of organization. Included here are the basic parts: the major cell groups (nuclei, cortical areas, and so on) and the major fiber tracts that interconnect them. The macrocircuitry of the brain is laid down during embryogenesis by a genetic program that has evolved in a unique way for each species. The brain of each species has a unique and characteristic macroarchitecture, just as the body as a whole has. This is how species were originally defined (Chapter 1). Every individual of a species is different, but those differences are within a narrow range compared to differences between species. In contrast, the microcircuitry of the brain is concerned with absolute numbers of neurons in a cell group, absolute numbers of axon collaterals and dendritic spines, absolute strength of particular synapses, and so on. The microcircuitry of every individual is different and changes dynamically throughout life.

In summary, each species has a characteristic brain macrocircuitry that is genetically hard-wired during development, and each individual within a species has a unique brain microcircuitry that changes dynamically throughout life.

LEARNING *Changing Synaptic Strength*

In the late nineteenth century Cajal discovered how nerve cells interact with one another in the adult brain: an axon terminal comes into contact or contiguity with a dendrite or cell body. Quickly he realized that learning might be explained by changes in the strength of these functional contacts (synapses), in a way analogous to how muscle cells become stronger with use or exercise. Tremendous progress has been made in the last several decades toward understanding the cellular and molecular underpinnings of changes in synaptic strength.

It is probably best to start with the observation that there are two fundamentally different classes of learning: associative and nonassociative. *Habituation* is an example of nonassociative learning. In many cases, when a particular stimulus is presented over and over again, the magnitude of the stereotyped response to that stimulus progressively decreases. Then, if there is a long gap of time before the stimulus is presented again, a response as large as the first one is elicited. This simple type of learning is common to amoebas and humans. In animals with a nervous system, habituation appears to involve a transient weakening of synaptic strength; the converse response, sensitization, appears to involve a transient strengthening of synapses.

Associative learning is different and at least requires a nervous system. Experimental psychologists have traditionally distinguished between two broad types of associative learning: classical and instrumental. Classical conditioning or learning was made famous by Ivan Pavlov, who won the first neuroscience-related Nobel Prize, in 1904. The principle here is that certain stimuli reliably elicit a stereotyped response. For example, the sight of food (an unconditioned stimulus) elicits salivation (an unconditioned response) in a dog. By contrast, the ringing of a bell normally does not elicit salivation. However, if the ringing of a bell is paired in time with the sight of food—specifically, if it is rung *just before* the sight of food—the dog learns (associates) to salivate when it hears a bell but does not see the food. The sound of the bell has become a conditioned stimulus that leads to a conditioned response. As discussed in Chapter 6, Richard Thompson and his colleagues have made great progress in clarifying the organization of neural circuits that mediate classical Pavlovian learning. A key site for synaptic plasticity has been identified in the deep cerebellar nuclei, although the exact chemistry remains to be clarified.

Instrumental conditioning or learning was made famous by B.F. Skinner of Harvard. In this case, an animal or person must do something actively and experience the consequences of the behavior. For example, a rat might eventually press a lever in its laboratory cage for the first time and unexpectedly receive a tasty morsel to eat. The animal quickly learns that pressing the lever delivers something good to eat. This is reinforcement learning. The subject does something and then receives as feedback positive (pleasant) or negative (unpleasant) reinforcement—a reward or a punishment. This association of pain or pleasure with the execution of behavioral acts is a powerful determinant of future behavior—that is, whether particular behavioral acts will be repeated or avoided. One of the main differences between classical and instrumental conditioning is that the former involves a passive situation: the animal or person is simply exposed to two stimuli (unconditioned and conditioned), and over the course of one or more trials a conditioned response is learned. In contrast, instrumental conditioning requires active participation of the subject, who must voluntarily initiate a behavioral act and then receive feedback about the results. The subject is instrumental in initiating the learning event.

The organization of neural networks that mediate instrumental leaning is not nearly as well-understood as that for classical learning. However, because voluntary initiation of behavior is critical for instrumental learning, it seems very likely that the critical site of synaptic plasticity is in the cerebrum (Chapter 8) rather than the cerebellum. Unfortunately, the link between positive and negative reinforcement systems (pain and pleasure systems, as discussed in Chapter 9) and cerebral synaptic plasticity remains obscure.

The best model for studying the chemistry of synaptic plasticity, and the possible role of modified gene expression associated with learning, is a phenomenon called *long-term potentiation* (LTP). It has been exploited most thoroughly in the hippocampal cortex, which seems to play an important role in learning spatial information—for example, about the environment during exploratory or foraging behavior. However, LTP is found in many other parts of the nervous system, including the sympathetic ganglia, where a form of it was discovered by M.G. Larrabee and D.W. Bronk in 1947. Work on the hippocampus began in 1973 when T.V.P. Bliss and T. Lømo basically showed that the postsynaptic response to an action potential is enhanced or potentiated if it is preceded by an appropriate burst of action potentials. In other words, under the right conditions involving the right cell types, postsynaptic responses can be greatly augmented by preceding patterns of action potentials. Short-lasting augmentation associated with sensitization, for example, had been known for some time. What made LTP unusual was its long duration. In intact rats, hippocampal LTP can last for at least months.

The chemical underpinnings of LTP have proven to be very complex and sometimes different in different classes of synapses. In fact, the actual long-lasting biochemical change ultimately responsible for enhanced synaptic transmission remains elusive. However, certain initial stages of the process are clear. For example, it seems certain that the biochemical changes are triggered by increased entry of Ca^{2+} ions into the postsynaptic compartment and that these ions enter through special glutamate-sensitive receptors (NMDA receptors) that only open when the postsynaptic membrane is depolarized (for example, by a train of action potentials).

The bottom line is that there are many mechanisms for changing the strength of synapses with use, and it is entirely possible that the efficacy of transmission at all synapses is subject to modification by use. These cellular mechanisms range from habituation and sensitization, through tetanic and posttetanic potentiation, to long-term potentiation and depression. They are electrophysiological measures of synaptic plasticity. It is important to end this section with the observation that LTP may also be accompanied by morphological changes. There is evidence to suggest that LTP is accompanied not only by changes in the shape of synapses (for example, larger postsynaptic densities, which imply more effective synaptic transmission) but also by an increase in the number of synapses, or at least synaptic densities. Thus, there is increasing evidence that at least some forms of learning are accompanied by changes in the brain's physical microcircuitry.

STRESS *Biochemical Switching*

Stress has been defined as any condition that perturbs bodily mechanisms from their normal equilibrium state. Curiously, a very good empirical definition of stress has turned out to be any stimulus or condition that elicits secretion of ACTH from the pituitary gland, and thus the release of glucocorticoid steroid hormones (for example, cortisol) from the cortex of the adrenal gland into the blood (see Fig. 6.14). There is an essentially infinite set of conditions that produce stress—from exposure to a hot or cold environment, to confrontation with a predator, to public speaking. Yet despite the fact that dealing with each one of these situations requires a unique, customized set of responses, they all share one feature: increased blood levels of glucocorticoid hormones. As the name implies, one of the important effects of these hormones is to raise blood levels of glucose, thus helping to supply more energy for reacting successfully to the stressful situation.

The hypothalamo-pituitary-adrenal axis is a classic example of negative feedback control in a neuroendocrine system (see Fig. 6.14). The basic idea is that high circulating levels of adrenal glucocorticoid hormones feed back on the hypothalamus to decrease the synthesis and release of CRH—the hypothalamic peptide hormone/neurotransmitter that secretes ACTH from the anterior pituitary gland—and low levels of circulating glucocorticoids have the opposite effect: they lead to increased synthesis and release of CRH. This arrangement serves to maintain relatively consistent levels of circulating glucocorticoids.

Glucocorticoids have two important features in the present context. First, they cross the blood–brain barrier and thus gain unimpeded access to the brain from the blood. Second, they have widespread effects on gene expression via nuclear glucocorticoid receptors, which bind to regulatory regions of DNA when occupied by hormone. After Wylie Vale and his colleagues identified CRH and raised antibodies to it in the early 1980s it became possible to examine experimentally the hypothalamic-pituitary-adrenal axis with immunohistochemical (and later on with hybridization histochemical) methods. On a very basic level it has been found that glucocorticoids exert a profound inhibitory effect on expression of the CRH gene (and on levels of CRH peptide in hypothalamic neuroendocrine CRH neurons) in the paraventricular nucleus. This was not surprising, but unexpectedly it was found that these CRH neurons express two additional neuropeptide genes when glucocorticoid negative feedback is removed; these are the genes for vasopressin and angiotensin II, both of which stimulate secretion of ACTH. Thus, when blood levels of glucocorticoids are chronically low, neuroendocrine CRH neurons synthesize three ACTH secretogogues, which act synergistically and thus very powerfully on ACTH release.

These results indicate that glucocorticoid hormones can dramatically alter the ratio of neuropeptides synthesized and shipped down the axon of individual neurons. However, the situation with CRH neuroendocrine motoneurons is much more remarkable than this. It is now known that this one cell type can express more than 10 different neurotransmitters, and immunohistochemical and hybridization histochemical methods have been used to show that each type of stress that an animal is exposed to produces a different ratio of these neurotransmitters within this neuronal cell type. In other words, the complement of peptide neurotransmitters found in CRH neuroendocrine motoneurons at any particular time is a function of the history of the animal—what stressors it has been exposed to over the course of the last several days. This result might not seem surprising in light of our earlier comment that each type of stressful condition requires a unique set of physiological and behavioral responses. However, the actual functional consequences of having different complements of neurotransmitters in a neuron at different times are much more difficult to determine.

CRH neuroendocrine motoneurons provide a fascinating model for testing predictions about the functional consequences of altered ratios of neurotransmitters within individual neurons. In addition to expressing more than a dozen potential neurotransmitters, the axon of these neurons does almost everything that an axon can do (Fig. 10.1). First, their main projection is to the median eminence, where their axon terminals release whatever complement of neurotransmitters are available into the portal circulation for delivery to the anterior pituitary. This is a hormonal function. In addition, some of the transmitters released in the median eminence bind to receptors on nearby axon terminals (presynaptic receptors). For example, CRH appears to inhibit the release of GnRH in the median eminence of

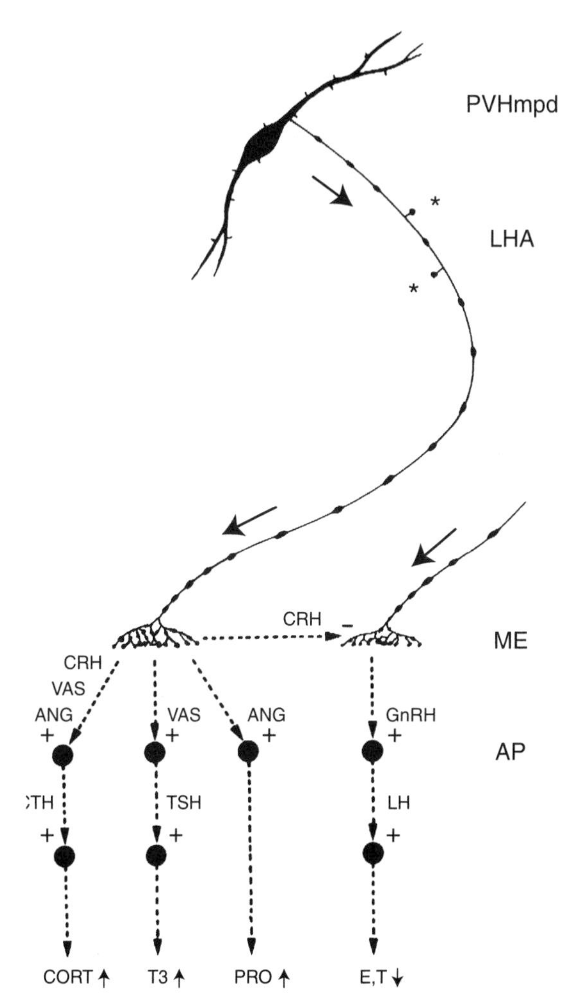

FIGURE 10.1 *This drawing shows the basic morphology of a typical CRH neuroendocrine motoneuron in the dorsal part of the medial parvicellular division of the hypothalamic paraventricular nucleus (PVHmpd). The parent axon ends as a spray of terminals (telodendron) in the external layer of the median eminence (ME), although boutons (synapses) of passage occur along the axon in the lateral hypothalamic area (asterisks, *; LHA). In the median eminence, neurotransmitters can have two actions. First, they can diffuse to nearby axon terminals with appropriate presynaptic receptors (say, for CRH) and exert paracrine effects. Second, they can enter the hypophysial portal circulation and be transported to the anterior pituitary gland (AP). In this way, three of the neuropeptides released by these neurons can directly or indirectly (via paracrine effects) influence the secretion of hormones from four cell types in the anterior pituitary.* Key: *ACTH, adrenocorticotropic hormone; ANG, angiotensin II; CORT, cortisol/corticosterone; CRH, corticotropin-releasing hormone; E, estrogen; GnRH, gonadotropin-releasing hormone; LH, luteinizing hormone; PRO, prolactin; T, testosterone; T3, thyroid hormone; TSH, thyroid-stimulating hormone; VAS, vasopressin. Adapted with permission of Elsevier Science from L.W. Swanson, Biöchemical switching in hypothalamic circuits mediating responses to stress,* Prog. Brain Res., *1991, vol. 87, p. 192.*

rats. This is a paracrine effect of CRH on GnRH-containing axon terminals with CRH receptors. And finally, on their way to the median eminence, CRH neuron axons generate boutons-of-passage within the lateral hypothalamic area. In all likelihood this represents a synaptic function of this CRH cell type. Thus, CRH neuroendocrine motoneurons are in a position to mediate synaptic, paracrine, and endocrine effects on different cell types in the lateral hypothalamic area, median eminence, and anterior pituitary, respectively.

We have suggested that this arrangement could result in the *biochemical switching* of information flow through an anatomically fixed circuit (Fig. 10.1). First let us examine the hormonal effects of CRH on the anterior pituitary. Under chronically high levels of circulating glucocorticoids (as in Cushing's disease), CRH neurons are making little or no vasopressin and angiotensin and only moderate amounts of CRH. Here, a given number of action potentials will release some CRH but little or no vasopressin or angiotensin. This will lead to modest release of ACTH and thus adrenal glucocorticoids. In contrast, under chronically low levels of circulating glucocorticoids (as in Addison's disease), high levels of CRH, vasopressin, and angiotensin II will be synthesized and relatively large amounts will be released by the same number of action potentials. This will lead to very high levels of ACTH release but will also lead to thyroid-stimulating hormone release because vasopressin is a secretagogue for this anterior pituitary hormone as well. In addition, prolactin will be secreted because angiotensin II is also a secretagogue for this hormone. Thus, under one hormonal condition CRH neurons will release ACTH, and under another condition they will release ACTH plus thyroid-stimulating hormone and prolactin.

But what about all of the other neurotransmitters synthesized by these CRH neurons? Two of them are of particular interest here: enkephalin and neurotensin. They are interesting because there appear to be no receptors for these peptides in the anterior pituitary. Instead, there are enkephalin receptors in the median eminence (presynaptic receptors on axon terminals) and neurotensin receptors in the lateral hypothalamic area. Thus, one could reasonably hypothesize that these CRH neurons synthesize a number of transmitters in part because there are different complements of receptors at the various sites where their neurotransmitters act. For example, neurotensin may exert postsynaptic effects on neurons in the lateral hypothalamic area but no effects in the anterior pituitary where receptors are lacking. The concept of biochemical switching of information flow in neural circuits is easy to appreciate in the lateral hypothalamic area. All things being equal, this synapse will not function if transcription of the neurotensin gene has been inhibited for any length of time, and only neurotensin receptors are expressed postsynaptically. Or put another, more likely, way: the efficacy of this synapse may depend on how expression of the neurotensin gene is regulated.

In the case we have been considering, a particular steroid hormone produces reversible effects on expression levels of a set of neuropeptide genes in a particular neuronal cell type. However, there is now a vast literature showing that other hormones, as well as neurotransmitters released from axon terminals, can regulate the expression of neurotransmitters and neurotransmitter receptors in many regions of the nervous system. Furthermore, each experimental manipulation of the animal seems to produce a unique "signature" pattern of gene expression changes in the nervous system. It turns out that there is constant, extensive biochemical plasticity in neural circuits. Exactly how this influences information processing ("computing") in these circuits remains largely unknown.

CYCLES *Circadian and Reproductive*

In Chapter 7 we discussed circadian and reproductive rhythms, and it is now clear that they are accompanied by changes in gene expression that is potentially related to the efficacy of synaptic transmission. The implication here, as in the preceding section, is that information processing in neural circuits is not simply a product of action potential patterns in the network. Information processing may also be influenced by changes in the availability of neurotransmission-related molecules as determined by altered levels of corresponding gene expression. For example, there is a clear circadian rhythm of CRH gene expression in paraventricular neuroendocrine motoneurons, and there are also clear circadian rhythms of neuropeptide gene expression in the suprachiasmatic nucleus itself (the primary endogenous circadian clock of the brain).

Another example of changing gene expression patterns under natural conditions involves the female reproductive cycle in rodents, where it has been examined most carefully. The approximately 4-day estrous cycle in female rats was discussed in Chapter 7, where it was pointed out that the animals go into heat once every 4 days, at the time of the cycle when ovulation takes place. This coordination of behavioral receptivity and ovulation maximizes chances of egg fertilization through sexual intercourse and is driven by a surge of estrogen controlled by the hypothalamus. It was pointed out that the estrogen surge produces a major shift in the female's behavior, from defending against the advances of males to actively soliciting a partner for mating. It takes about 8 hours for the effects of estrogen to be manifest, and they are almost certainly due to effects of the steroid hormone on some aspect(s) of neurotransmission-related gene expression in the sexually dimorphic circuit of the forebrain.

Exactly how estrogen produces these specific changes in behavior is not known. However, clear examples of estrogen effects on neuropeptide gene expression in the sexually dimorphic circuit over the course of the estrous cycle have been demonstrated. For example, substance P and cholecystokinin are coexpressed in neurons of three interconnected parts of the circuit (medial amygdala, BST, and medial preoptic nucleus). Over the course of the estrous cycle, substance P levels stay constant in these neurons, whereas cholecystokinin only becomes detectable on the day of estrus. In other words, ratios of coexpressed neurotransmitters change dramatically within individual neurons of the sexually dimorphic circuit during the course of the female reproductive cycle due to changing levels of estrogen in the blood. This is another example of potential biochemical switching or biasing of information flow through a functionally specific neural network.

DAMAGE REPAIR *Regrowth*

Perhaps the most dramatic examples of dynamic architecture in the adult mammalian nervous system involve responses to damage and disease. This is another vast topic that we simply broach here because of its critical importance from a medical point of view. First and foremost, there is a basic difference between damage repair in the peripheral and in the central nervous systems. When a peripheral nerve is cut, the distal end inevitably degenerates, of course (Appendix C). However, if a skillful surgeon unites the severed halves of the nerve carefully, the intact (central) stump of the nerve can regrow along the old pathway to the original innervation fields, and sensation can be restored. This regeneration is more successful the closer to the periphery the cut has been made.

The situation is quite different in the central nervous system, where lesions, destruction, or death of neurons rarely lead to any significant regeneration of previously intact circuits. One reason for this unfortunate situation is the extreme complexity of the brain. Regrowing axons would have to navigate an unbelievably complex labyrinth of neural tissue to find their original targets. The other impediment may prove to be more tractable. When central neural tissue is damaged, a "glial scar" forms in the region. It is produced by the massive proliferation of supporting cells (primarily astrocytes) that remove damaged tissue through phagocytosis. In addition, however, cells of the glial scar may secrete factors that inhibit the growth of axons.

It has been known since the nineteenth century that damaged neurons in the brain attempt to regrow their axons. Unfortunately, these axonal sprouts typically do not grow very far. Somewhat more success has been obtained by transplanting certain neurons into the damaged brain. For example, certain transplanted aminergic neurons (Chapter

7) can send new axons rather large distances through the brain. In all these cases, however, it is important to determine whether newly generated axons in the adult brain establish correct or incorrect synaptic relationships. Do they reestablish circuitry that was damaged, or do they establish new connections that in essence form aberrant circuitry? The latter situation could well be worse than no repair at all. This is one of the basic conundrums of experimental neurology—how to repair damaged neural circuits without creating more harm than good, more side-effects than benefits. The structural complexity of the central nervous system is a formidable opponent.

Perhaps the greatest hope for repairing the damaged adult nervous system lies in understanding the cell and molecular biology of nervous system development. The goal here is to take advantage of the molecular mechanisms responsible for building neural circuits in the embryo. Perhaps they can be reinitiated or mimicked in the adult to rebuild or repair damaged circuits.

READINGS FOR CHAPTER 10

Larrabee, M.G., and Bronk, D.W., Prolonged facilitation of synaptic excitation in sympathetic ganglia. *J. Neurophysiol.* 10:139–154, 1947.

Kandel, E.R., Schwartz, J.H., and Jessell, T.M. *Principles of Neural Science*, fourth edition. McGraw-Hill: New York, 1999. It has a good introduction to the neuroscience of learning and damage repair.

Sawchenko, P.E., Li, H.Y., and Ericsson, A. Circuits and mechanisms governing hypothalamic responses to stress: a tale of two paradigms. *Prog. Brain Res.* 122:61–78, 2000.

Simerly, R.B., Young, B.J., Capozza, M.A., and Swanson, L.W. Estrogen differentially regulates neuropeptide gene expression in a sexually dimorphic olfactory pathway. *Proc. Natl. Acad. Sci. USA* 86:4766–4770, 1989.

Swanson, L.W. Neuropeptides: New vistas on synaptic transmission. *Trends Neurosci.* 6:294–295, 1983.

Swanson, L.W. Histochemical contributions to the understanding of neuronal phenotypes and information flow through neural circuits: the polytransmitter hypothesis. In: *Molecular Mechanisms of Neuronal Communication*, K. Fuxe, T. Hökfelt, L. Olson, D. Ottoson, A. Dahlström, and A. Björklund (eds.). Pergamon Press: New York, 1996, pp. 15–27.

Zigmond, M.J., Bloom, F.E., Landis, S.C., Roberts, J.L., and Squire, L.R. (eds.) *Fundamental Neuroscience*. Academic Press: San Diego, 1999. This textbook also has a good introduction to learning and damage repair.

11

Gene Networks

Relationship to Neural Networks

> It is not a little remarkable that what is definitely known regarding the special functions of the nervous system has been ascertained within the last thirty years.
>
> —BRITISH AND FOREIGN MEDICAL REVIEW (1840)

> Although I disapprove the ludicrous "scientific" attitude displayed by many "molecular biologists" and believe that their extravagant shenanigans deserve a full measure of ridicule, I nevertheless consider their specialty to be an important field of biology.
>
> —HARTWIG KUHLENBECK (1973)

So far we have outlined the cellular composition of the nervous system, the spatial relationships of its basic parts, and the network arrangement of its four basic functional systems. This is classical neuroscience. Like any organ the brain has a regional architecture and a number of functional systems, in this case motor, cognitive, state control, and sensory. The global architecture of the vertebrate nervous system seems to be based on the principle of segmentation, with particular neuromere segments differentiating to different extents in different groups of animals. Perhaps the most critical factor in this differentiation is the overall body plan of a particular animal group. The fundamental building block or component of all neural networks is the neuron, which has the same basic cell biology in all animals with a nervous system. What varies between species is the way neurons are used to build the neuromeres—or, alternatively, the way neurons are used to construct neural circuits or networks within the system as a whole.

But there is a seemingly very different way of looking at the function of the nervous system—in terms of chemical systems and, at a more basic level, gene expression patterns. This point of view is concerned with how drugs act on the brain and how altered patterns of gene expression influence nervous system structure and function. The basic conundrum here is that very often a drug acts, or the expression of a gene is altered, in a complex way that cuts across multiple functional systems. For example, acetylcholine is a neurotransmitter in somatic motoneurons of the spinal cord, in parasympathetic ganglia of the visceral organs, in magnocellular neurons of the basal forebrain, and in cerebral cortical interneurons. Therefore, the enzyme that synthesizes acetylcholine is expressed in specific cell groups that are nevertheless parts of quite different functional systems. Expression of the gene for this enzyme is regulated in a very neuron-specific way, but the cell groups involved are not related in any known way to a particular functional system. In other words, quite possibly there is no obligatory relationship between functional systems in the brain and global patterns of gene expression.

If this is true, then certain basic conclusions follow. For example, if there is no relationship between functional neural systems and gene expression systems or patterns, then drugs will typically act on multiple functional systems, and the altered expression of any single gene will typically occur in multiple functional systems. If this is true, it also follows that gene expression patterns cannot reveal anything about the organization of functional neural systems.

The important point here is that the functional and pharmacologic/genomic systems of the brain are equally important in their own right, whether or not they are causally interrelated. The practical implications for no relationship are that, generally speaking, individual drugs will have multiple side effects (effects on multiple systems), and the effects of genetically engineering the expression of a particular gene will be complex and multifunctional. At the moment, examples of highly specific drug action or highly localized gene expression in one particular functional system are certainly known but are exceptions to the rule.

Today, the genome has at least in principle been sequenced—almost exactly 50 years after the molecular structure of DNA was elucidated by James Watson and Francis Crick at Cambridge. Of course the task of decoding the sequence has now begun in earnest, and the results of this enterprise ultimately will resolve the problem of if and how neural networks are related to gene networks. Nevertheless, the sequencing itself has major theoretical and technical implications. On the theoretical, computational, or modeling side, we now know that there are on the order of 30,000 to 60,000 genes in the mammalian genome (with well over half thought to be expressed in the brain). This knowledge is fundamentally important because we now have boundary conditions on the problem of gene network complexity. In principle we can know what all of the genes are—what all of the players are—and we have begun to classify them functionally. It is only a matter of time before the function of all the genes will be known, and they will be classified in an orderly way. On the technical side of the coin, this knowledge allows us in principle to measure how the entire genome is expressed over time in any particular part of the brain, and under any conditions, we are interested in.

One goal of molecular biology is to understand how the network of 30,000 to 60,000 genes in the chromosomes of each cell is regulated as a whole over the course of time. It now seems certain that expression in the gene network is modulated by the combinatorial action of an exceptionally rich set of regulatory (transcription) factors. The difficulty will come in trying to determine experimentally the kinetics of this regulation. In the end, however, we are faced with a problem in complex systems analysis.

The architecture of the brain is also, as we have seen, a problem in complex systems analysis. By coincidence, the complexity of neural and gene networks may be on roughly the same order of magnitude. It has been estimated that there are in round numbers about

50,000 major connections or pathways that form the macrocircuitry of the central nervous system. It is hard even to imagine at this point in time how two different systems as complex as this could be compared in a systematic way.

In the twenty-first century neuroscience will be transformed by molecular biology in ways that would be foolish even to speculate about. We are in the early stages of a revolution as transforming as the introduction of the cell theory in the middle of the nineteenth century. If history is any guide, we can expect that the fundamental contributions of molecular biology to the architecture of the brain will come from two sources. One source will involve comparative studies of much simpler organisms, and the other will involve experimental analysis of early mammalian development, when neural macrocircuitry literally is being constructed by a genetic blueprint or program that remains to be decoded or reverse engineered. It will be exciting to see whether molecular biology ends up basically confirming 2500 years of thinking about the architectural plan of the brain, whether it provides a radically different interpretation, or whether it proves to be irrelevant.

READINGS FOR CHAPTER 11

Brenner, S. Theoretical biology in the third millennium. *Phil. Trans. Roy. Soc. B* 354:1963–1965, 1999.

Davidson, E.H. *Genomic Regulatory Systems: Development and Evolution.* Academic Press: San Diego, 2001.

Evans, R.M., Swanson, L., and Rosenfeld, M.G. (1985) Creation of transgenic animals to study development and as models for human disease. *Rec. Prog. Hor. Res.* 41:317–337, 1985.

Leighton, P.A., Mitchell, K.J., Goodrich, L.V., Lu, X., Pinson, K., Scherz, P., Skames, W.C., and Tessier-Lavigne, M. Defining brain wiring patterns and mechanisms through gene trapping in mice. *Nature* 410:174–179, 2001.

Wade, N. *Life Script: How the Human Genome Discoveries Will Transform Medicine and Enhance Your Health.* Simon and Schuster: New York, 2001.

Watson, J.D., and Crick, F.H.C. A structure for deoxyribose nucleic acid. *Nature* 171:737–738, 1953.

Appendix A

Describing Position in the Animal Body

No art in the world can render on paper the microscopic views as the eye sees them.
BENEDICT STILLING (1856)

It would seem to go without saying that anatomy is critically dependent on the unambiguous description of physical relationships. Therefore it is both surprising and confusing to find how difficult it is to read the neuroanatomical literature. It is bad enough, as we shall see in Appendix B, that names for the parts are not standardized. But it is very disconcerting that even words used to describe position within the central nervous system are often ambiguous. For example, whereas in geography the meanings of *north*, *south*, *east*, and *west* are universally understood, the terms *anterior* and *posterior* often have contradictory meanings in embryology and human gross anatomy.

Why is there such confusion about describing position or location in neuroanatomy? Many factors are undoubtedly involved, but the most important is probably tradition. From classical Greek times through the end of the eighteenth century the overwhelming interest was in the structure of the human body. Unfortunately for descriptive anatomy, we are somewhat unusual in our typical bipedal mode of locomotion, and this upright posture as compared to quadrupeds, snakes, and fish has led to the development and use of an idiosyncratic terminology for human anatomy. The obvious long-term solution has been emphasized in the earlier chapters of this book and falls back on the time-honored comparative and embryological approaches. Positional descriptors are most clear and unambiguous when they refer to the idealized relationships observed in the "typical vertebrate body plan" (Fig. 4.2), and in the "straightened-out embryo" (Figs. 4.6, 4.10). One beauty of this approach is that the same simple, clear set of positional descriptors can be applied to all vertebrates (and all bilaterally symmetrical invertebrates as well): rostrocaudal, dorsoventral, and mediolateral (Fig. A.1, lateral and dorsal views at the top). But for now, when reading the literature one needs to infer positional meaning from context and a little background knowledge. Readers interested in pursuing this topic in depth should consult P.L. Williams 1995. The following is an introductory overview.

It is important to know that the vertebrate body is described in terms of three perpendicular axes along with three corresponding planes. The axes (rostrocaudal, dorsoventral, and mediolateral) are a little like the north–south and east–west lines on a compass. They are perpendicular to one another. One can travel a certain distance north or south, just as one can progress a certain distance rostral or caudal. Because the body is a three-dimensional object, rather than a surface (of the earth), three rather than two axes or cardinal directions are needed. They correspond to the x, y, and z axes of Cartesian geometry.

Now imagine an adult human standing up (Fig. A.1, lower right). In humans, the "back" of the body, the dorsum, is traditionally referred to as *posterior*, whereas the "front" or belly, the ventrum, is traditionally referred to as *anterior*. So in the spinal cord gray matter, for example, one typically refers to anterior horns in humans and to ventral horns in animals. This can be very confusing because some embryologists insist on referring to the rostrocaudal axis as the anterior–posterior axis.

In human anatomy one also traditionally refers to structures toward the head as *superior* and those toward the feet as *inferior*. This convention inspired the names of certain structures in the human

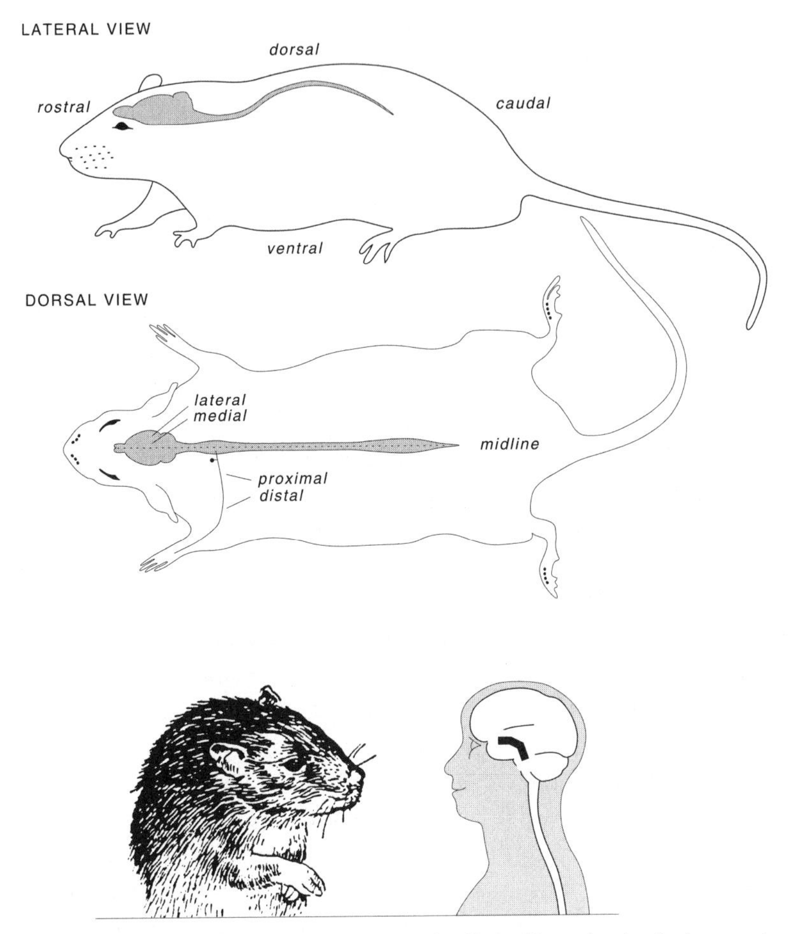

FIGURE A.1 *These drawings illustrate how to describe position or location in the rat and human. Reproduced with permission from L.W. Swanson,* Brain Maps: Structure of the Rat Brain, *second edition (Elsevier Science: Amsterdam, 1998–1999, p. 9).*

brain—a good example we will mention shortly is the superior and inferior colliculi of the midbrain tectum—that make no sense in comparative neuroanatomy.

In comparative anatomy there are three standard planes that are perpendicular to one another, and that cut through the body or unfolded embryo: a transverse plane and two longitudinal planes (sagittal and horizontal). In Cartesian geometry they would be equivalent to x, y, and z planes. The sagittal plane is the same in all bilateral animals, including humans. It is the longitudinal plane that cuts the body into right and left pieces when viewed from above (Fig. A.1, dorsal view). The midsagittal plane of course runs down the midline and cuts the body into right and left halves. Proceeding laterally from the midline creates parasagittal planes. Now the confusion begins. In humans, a series of planes proceeding from superior to inferior (in the standing subject) is referred to as *horizontal*, which in context makes perfect sense. However, in comparative anatomy, where one most commonly deals with quadrupeds, fish, and snakes, the horizontal plane has a completely different meaning. Here it is the longitudinal plane that is perpendicular to the sagittal plane, and so it also has a rostrocaudal orientation. In animals the horizontal plane is parallel not perpendicular to the back (dorsum). Because of this, the third standard plane—transverse, coronal, or frontal—must also have a fundamentally different meaning in human and in comparative anatomy. In human anatomy the coronal plane is parallel to the long axis of the body and is at right angles (orthogonal) to the sagittal plane. In comparative anatomy, the coronal plane is truly transverse to the longitudinal axis of the body.

It is hard to predict how long this terminological standoff will persist. On one side are the comparative anatomists who are dealing with evolving general principles; on the other side is the medical community, which is very influential and conservative—and understandably anthropomorphic.

But the fundamental problem is of course more difficult because the longitudinal axis of the animal body and thus the central nervous system typically is not straight. The longitudinal axis of all vertebrate embryos undergoes a complex change in shape during development (Fig. A.2), and even the longitudinal axis of the adult rat is not straight (Fig. A.1, lateral view, and Fig. A.2). However, there is a critical feature of the human body, in particular of the human brain, that is responsible for serious confusion. There is an approx-

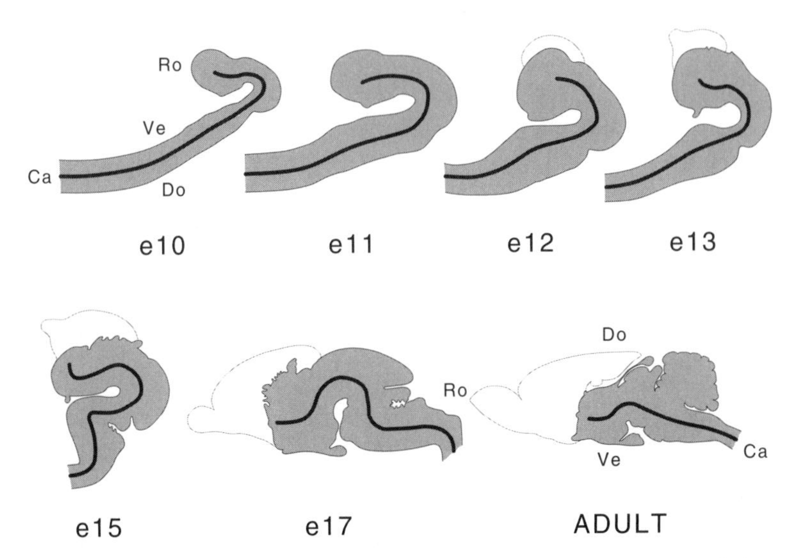

FIGURE A.2 *During the course of embryogenesis the longitudinal axis (thick black line) of the neural tube undergoes major changes in shape.* Key: *Ca, caudal; Do, dorsal; e10–e17, embryonic days 10–17; Ro, rostral; Ve, ventral. Adapted with permission from G. Alvarez-Bolado, and L.W. Swanson,* Developmental Brain Maps: Structure of the Embryonic Rat Brain *(Elsevier Science: Amsterdam, 1996, p. 29).*

imately 90-degree bend in the longitudinal axis of the human brain that occurs in the midbrain region (Fig. A.1, lower right).

Basically, the longitudinal axis of the spinal cord and hindbrain is vertical in the standing human, whereas the longitudinal axis of the forebrain is horizontal. This is "because" the standing human is looking forward: the face is parallel to the belly instead of perpendicular as in the rat (Fig. A.1, lateral view at top). The practical consequence of this arrangement is that if a series of planes is drawn, or a series of histological sections is cut, through the human brain, they will start off "in front"(within the frontal pole of the forebrain) in a plane transverse to the longitudinal axis of the central nervous system, but then "in back" (in the hindbrain and spinal cord, and in the "back" half of the cerebrum and in the whole cerebellum) they will be parallel to the longitudinal axis. In comparative anatomical terms, they will be frontal sections rostrally, horizontal sections caudally, and a series of intermediate planes in between (where the superior and inferior colliculi lie).

Based on strict physical relationships, it seems obvious that there can be no logically consistent way to apply a strict Cartesian coordinate system (with three standard, perpendicular planes) to the general vertebrate brain, or even to any individual species, if rostrocaudal actually refers to the longitudinal axis. Logically, the general solution would seem to lie in the topological relationships of parts, rather than in their physical relationships, which become distorted in unique ways in different groups of animals during embryogenesis. Such an approach is not yet common, although it has been adopted in this book.

Really effective description avoids slang. This is why the use in structural neuroscience of everyday terms like "in front of," "behind," "on top of," and "under" should be strictly avoided. There is enough confusion about the meaning of the technical terms discussed here. It is amazing how often the meaning of these common terms is very unclear, and thus textual descriptions are ambiguous, when reading the older neuroanatomical literature, and the same will eventually happen when they are used now.

READINGS FOR APPENDIX A

Williams, P.L. (ed.) *Gray's Anatomy*, thirty-eighth (British) edition. Churchill Livingstone: Edinburgh, 1995, pp. xv–xvii.

Appendix B

Naming and Classifying Nervous System Parts

Among the various parts of the animated Body, which are subject to Anatomical disposition, none is presumed to be easier or better known than the Brain, yet in the meantime, there is none less or more imperfectly understood.

—THOMAS WILLIS (1681)

The principal reason for the frequent disputes over terminology is not so much about whether a new term muddles Greek with Latin. It is really about whether the term is biased toward their theory rather than ours.

—MARCUS JACOBSON (1993)

In Appendix A we dealt with the widespread confusion about how to describe position or location within the central nervous system. We go on now to discuss the more substantive yet even greater problems of inconsistent neuroanatomical nomenclature, as well as the lack of rigorous classification schemes for the parts of the nervous system.

There is universal consternation when reading the highly international neuroanatomical literature about the meaning of terms that describe parts of the central nervous system. This is a problem that goes right back to the beginnings of the science in ancient Greece, but its magnitude can be glimpsed from the fact that over a century ago some 9000 terms had already been used to describe about 500 parts of the brain. According to Burt Wilder's presidential address to the American Neurological Association in 1885, there were in round numbers 2600 terms in Latin, 1300 in English, 2400 in German, 1800 in French, and 900 in Italian and Spanish. It is sobering to think how many terms may have been added since then. The number is entirely unknown but must be enormous.

Unfortunately the problem is not with synonyms. There are indeed many synonyms, but they are relatively easy to deal with. The real problems come when the same term is used for different structures, when there are varying interpretations about the borders of particular structures and how those structures might be subdivided, and when authors use terms without defining them.

These have always been serious problems because they cause ambiguity in the data presented in the literature. As a result, readers can misinterpret data, or they may simply ignore a body of important data because it is difficult to interpret. However, they are even more serious now that massive electronic neuroscience databases are on the horizon.

When all is said and done, the reason for this crippling situation is simple. It is not because anatomists are lazier, sloppier, or less critical than other scientists. After all, there is essentially no controversy about how to name the bones, muscles, and blood vessels. This has been settled for hundreds of years. The confusion in neuroanatomical nomenclature is due primarily to the fact that, unlike the skeletomotor and cardiovascular systems, we do not understand the basic organization of the nervous system. Stated another way, there are many, many areas of genuine controversy about the neuroanatomy of brain parts that await more data for resolution. After all, we are attempting to analyze an organ that is orders of magnitude more complex than any other part of the body, and from a realistic point of view we are only at the very initial stages of this analysis.

History has repeatedly shown that attempts to enforce a rigid nomenclature on brain regional anatomy are doomed to failure, and for good reason. Our understanding of brain architecture is evolving quickly, and it is entirely possible that most of the nomenclature popular at the moment will not be relevant a century from now. Neuroanatomy has a specialized technical language of its own, and like any language it evolves. Particular terms are preferred when they are found useful by most people in the field. Because of the power of words in reifying concepts, it is a major mistake at this point in time to consider trying to enforce a "standard" nomenclature for the parts of the central nervous system. This can only impede progress in trying to understand the actual structure of the brain, as free as possible of preconceived biases.

This is not to say that all neuroanatomical terms are equally valid or that there is not a great deal that can be done to clarify neuroanatomical nomenclature. The single most important thing neuroscientists can do in this realm is to define the anatomical terms they use and explain why they use the ones they do to the exclusion of others. If new terms are introduced, they should be carefully defined with respect to existing terms, and reasons for introducing the new term should be given. Most readers will probably be appalled at the trivial nature of this suggestion. However, it is even more appalling how infrequently neuroanatomical terms are defined and, conversely, how often their meaning is unclear. To reiterate: this ambiguity is not due to the use of synonyms, but it arises from differing interpretations of brain structure. The same word often has different meaning (structural interpretation) to different authors. The practical problem is this. When a neuroanatomical term is used in a specific paper, what does it mean to the author? There are almost always differing, critically unresolved views in the literature. This is the current state of neuroanatomy.

Many psychologists have argued that there is a natural tendency for the human mind to classify—it can't be helped. Thus, it comes as no surprise that there is a long history of attempts to classify the parts of the brain, although this went out of fashion in the latter half of the twentieth century. It stands to reason that classification schemes can only be as good as the data they are based on, and this may explain why there has been little interest in the topic lately. There is so much ambiguity in the literature, so many different interpreta-

- 1. Central Nervous System
  - 1.1 Brain (encephalon)
    - 1.1.1. Cerebrum (endbrain)
      - 1.1.1.1. Cerebral cortex
        - 1.1.1.1.1. Cortical plate (layers 1-6)
          - 1.1.1.1.1.1. cingulate region
          - 1.1.1.1.1.2. frontal region
          - 1.1.1.1.1.3. hippocampal formation
          - 1.1.1.1.1.4. insular region
          - 1.1.1.1.1.5. occipital region
          - 1.1.1.1.1.6. parietal region
          - 1.1.1.1.1.7. prefrontal region
          - 1.1.1.1.1.8. rhinal region
          - 1.1.1.1.1.9. temporal region
        - 1.1.1.1.2. Cortical subplate (layer 7)
      - 1.1.1.2 Cerebral nuclei (basal ganglia)
        - 1.1.1.2.1. Striatum
        - 1.1.1.2.2. Pallidum
    - 1.1.2. Cerebellum
      - 1.1.2.1. Cerebellar cortex
        - 1.1.2.1.1. Anterior lobe
        - 1.1.2.1.2. Posterior lobe
        - 1.1.2.1.3. Floculonodular lobe
      - 1.1.2.2. Cerebellar nuclei
    - 1.1.3. Brainstem
      - 1.1.3.1. Interbrain
        - 1.1.3.1.1. Epithalamus
        - 1.1.3.1.2. Dorsal thalamus
        - 1.1.3.1.3. Ventral thalamus
        - 1.1.3.1.4. Hypohalamus
      - 1.1.3.2. Midbrain
        - 1.1.3.2.1. Tectum
        - 1.1.3.2.2. Tegmentum
        - 1.1.2.2.3. Pretectal region
      - 1.1.3.3. Hindbrain
        - 1.1.3.3.1. Pons
        - 1.1.3.3.2. Medulla
  - 1.2. Spinal cord
    - 1.2.1. Cervical level
    - 1.2.2. Thoracic level
    - 1.2.3. Lumbar level
    - 1.2.4. Sacral level
    - 1.2.5. Coccygeal level

FIGURE A.3 *This is one scheme for a basic taxonomy of central nervous system parts. Adapted with permission from L.W. Swanson,* Brain Maps: Structure of the Rat Brain, *second edition (Elsevier Science: Amsterdam, 1998–1999, p. 194).*

tions, that synthetic approaches have disappeared at the expense of reductionistic focusing on narrower and narrower problems. However, in the last 25 years the reductionistic approach has produced vast amounts of neuroanatomical data that are much more reliable than ever before, so the time may be ripe to revisit this problem.

Classification or taxonomy is an important, synthetic method of analysis. Look how influential Linnaeus's binomial nomenclature scheme was for biology or how seminal Mendeleyev's periodic table of the elements was for chemistry. However, one must always bear in mind that any classification scheme inevitably reflects the biases of the author, just as the scheme inevitably imparts biases to those who study it. As a result, it is wise to subject any and all classification schemes to intense, ongoing scrutiny at the level of basic organizing principles.

This is especially true of schemes to classify the parts of the central nervous system. There are a number of essentially different ways of grouping central nervous system parts based, for example, on adult human regional anatomy, embryology and neuromeres, comparative and evolutionary neuroanatomy, and gene expression patterns (the genomic approach)—not to mention differing views within each of these broad categories. We have outlined one classification scheme in this book (see Figs. 4.15 and 4.17, and Fig. A.3), and have presented a detailed account elsewhere (see Swanson 1998–1999). However, it must be admitted that there is no irrefutable evidence for this, or any other, taxonomy of brain parts. It is presented as a model to stimulate further experimental work and the formulation of alternative schemes. And the same limitation applies to the taxonomy of functional neural systems presented in Chapters 5 to 9. At the moment, it seems to be the only modern global classification or model of brain network organization available. It begs replacement with a better one.

READINGS FOR APPENDIX B

Anthoney, T.R. *Neuroanatomy and the Neurologic Exam: A Thesaurus of Synonyms, Similar-Sounding Non-Synonyms, and Terms of Variable Meaning.* CRC Press: Boca Raton, Fla., 1994.

Eycleshymer, A.C. *Anatomical Names, Especially the Basle Nomina Anatomica ("BNA")*. William Wood: New York, 1917.

Swanson, L.W. *Brain Maps: Structure of the Rat Brain: A Laboratory Guide with Printed and Electronic Templates for Data, Models and Schematics*, second edition, with double CD-ROM. Elsevier: Amsterdam, 1998–1999, pp. 38–42.

Swanson, L.W. What is the brain? *Trends Neurosci.* 23:519–527, 2000. This is a short history of how the major parts of the brain have been named and classified.

Wilder, B.G. Paronymy versus heteronymy as neuronymic principles. *J. Nerv. Ment. Dis.* 12:1–21, 1885.

Appendix C

Methods for Analyzing Brain Architecture

> To examine each part [of the brain] thoroughly requires so much time and such application of mind that it would be necessary to give up all other labors and all other considerations on that particular task.
>
> —NICOLAUS STENO (1669)

> As long as our brain is a mystery, the universe, the reflection of the structure of the brain, will also be a mystery.
>
> —SANTIAGO RAMÓN Y CAJAL (1921)

> In the nervous system, we physiologists are more dependent upon what the anatomists tell us than we are anywhere else.
>
> —SIR JOHN ECCLES (1958)

Methods for analyzing brain structure divide broadly into two great classes. The oldest deals with regional anatomy—what one can see with the naked eye by dissecting the brain with knife, scraper, and probe. This approach is actually blossoming today thanks to exciting new technologies—for example, functional brain imaging, where dissection is carried out algorithmically with computer graphics. However, around the middle of the nineteenth century, the regional anatomy approach was supplemented and for many years largely overshadowed by revolutionary histological methods that allowed examination of neural tissue under the microscope, with orders of magnitude greater resolution. Thus began the era of cellular neuroscience that on the structural side has two branches: normal and experimental. Normal neurohistology deals with the microscopic appearance of neural tissue that has not been subjected to experimental manipulation, such as the production of lesions or the placement of tracer injections. All scientific techniques have advantages and disadvantages, and it is critical to understand what they are. No problem can be solved unequivocally with a single technique. The strongest argument for any position always comes from independent verification with independent methods.

For understanding the three-dimensional architecture of the brain as an organ, nothing remotely compares to personal dissection. The human brain, and the brains of animals like sheep and cows, are large objects, and a truly remarkable amount of structural organization can be observed by careful dissection (easily on the order of 500 major parts). It is probably not possible to obtain a reasonable appreciation for the structure of the brain as an organ strictly from the examination of histological sections (even a complete series) or from artistic renderings. All of the major differentiations of gray matter, and all of the major fiber tracts, can be examined with dissection, along with their fundamental shapes and topographic relationships. This is regional anatomy or architecture, and it is analogous to studying the distribution of land and water masses on a globe. It provides essential orientation for more detailed examination and description. The major limitation of this approach, obviously, is that it does not provide cellular resolution. It is not possible to determine the organization of neural circuitry with gross dissection (and by extension functional imaging methods, which actually have less resolution in common practice than naked eye examination of the brain).

Although the microscope was invented in the seventeenth century, virtually everything observed in the nervous system was artifactual until the 1820s when lenses that corrected serious spherical and chromatic aberrations began to be perfected in Germany. By the 1840s individual nerve fibers and neuronal cell bodies had been observed under the microscope, in a variety of animals, and it was then that Benedict Stilling began his unparalleled examination of the human brainstem, cerebellum, and spinal cord. In this work, which was carried out over a period of more than 20 years, Stilling examined under the microscope serial sections of this material cut in all three planes of section and described his results in a monumental series of books. Although no histological stains for neural tissue had yet been developed, he was able to see many neuronal cell groups for the first time. For example, he discovered most of the cranial nerve nuclei, as well as other major cellular features of the brainstem, cerebellum, and spinal cord.

In 1858 Joseph Gerlach introduced the first stain of any value for neurohistological material: carmine. It had a selective affinity for certain tissue features, especially the cell nucleus, so that cell bodies were easier to observe in brain tissue sections. Considerably better stains for neuronal cell bodies were not introduced until 1894 when Franz Nissl perfected the use of basic aniline dyes, which we now know stain nucleic acids, both in the cell nucleus and in the ribosomes of the cytoplasmic endoplasmic reticulum. This method remains a standard today. On the other side of the coin, Carl Weigert introduced in 1882 a stain for myelinated fiber tracts that is still in use today, and about a decade later Santiago Ramón y Cajal and Max Bielschowsky introduced reduced silver methods for staining axons themselves. Variations on the Weigert, Nissl, Cajal, and Bielschowsky methods provided a wealth of information about the general distribution of neuronal cell bodies and fiber tracts in the brain. However, they did not reveal the full morphology of individual neurons or the organization of neural circuits.

Three other normal histological approaches were indispensable. One was introduced by Camillo Golgi in 1873—the famous and revolutionary silver dichromate "black reaction" that for reasons still mysterious impregnates randomly about 1% of the neurons in a tissue section and impregnates them completely—axon, cell body, and all dendrites. Golgi gave the first adequate description of axon collaterals with this method, and Cajal went on to show how neurons contact one another in all regions of the adult brain. Cajal's work remains the cornerstone of our understanding of the cellular architecture of neural circuitry. In 1886 Paul Ehrlich introduced an entirely independent way to stain individual neurons completely, using methylene blue, and today neurons can be filled with markers using micropipettes that also record electrophysiological activity (and potentially obtain samples of intracellular content for molecular analysis). Together, these methods have been invaluable in determining the architecture of local circuitry. Until the last decade or so they have been much less useful in characterizing the long projections between neuronal cell groups.

The second major type of normal method is referred to as *histochemical.* Here chemical reactions are carried out on tissue sections, and the sites of these reactions are labeled in one way or another so that they can be observed under the microscope. For example, with this approach it is possible to determine the distribution cell by cell of neurotransmitters and their receptors. Today the most powerful histochemical techniques use antibodies to localize virtually any antigen of interest (immunohistochemistry) and complimentary strands of nucleic acids to localize specific mRNAs (in situ hybridization, or hybridization histochemistry).

The final type of normal method was introduced in the 1950s—electron microscopy. It provided about three orders of magnitude greater resolution (from about 1 μm with light microscopy), so that for the first time the structure of synapses could be observed, along with that of the myelin sheath and many of the intracellular organelles.

Now we come to the so-called experimental neuroanatomical methods for analyzing the structure of neural circuits. The incredibly complex meshwork of interconnections associated with neural circuits has proven impossible to analyze reliably without experimental pathway tracing methods. Experimental methods began with August Waller's demonstration in 1850–1851 that when a nerve is cut, the distal segment invariably degenerates. Ludwick Türck immediately extended this approach in a brilliant way by making lesions in the spinal cord and observing the distribution of Wallerian "secondary degeneration" in the descending tracts from the brain.

In the 1880s Carl Weigert attempted to trace pathways in the brain itself by making lesions and then observing the disappearance of fiber tracts with his myelin stain. Although a few things were discovered, it proved exceptionally difficult if not impossible to trace the loss of small tracts through the immense thicket of myelinated fibers distributed throughout the brain and spinal cord.

This problem was solved by V. Marchi and G. Algeri who introduced in 1885 a method for staining selectively degenerating myelin by itself, against a clear background of intact myelin. This approach has the obvious limitations that it does not reveal unmyelinated tracts, or the unmyelinated terminal regions of axons. These problems were not solved until the 1950s when W.J.H. Nauta and L.F. Ryan introduced the first selective stain for degenerating axons themselves.

The Marchi and Nauta methods are based on the phenomenon of anterograde (Wallerian) axonal degeneration. Obviously, this approach relies on the interruption of fibers-of-passage (or the neuronal cell bodies), and this is its greatest limitation. Very often the origin of fibers-of-passage was not known, and it was also common for fibers-of-passage of unknown origin to pass through the region of lesioned cell bodies. Thus, either data from lesion experiments were uninterpretible, or false positive results were obtained.

These problems were beautifully solved beginning in the early 1970s by taking advantage of normal physiological processes in neurons, most notably fast intra-axonal transport mechanisms. The first really successful method was based on the uptake of radiolabeled amino acids microinjected into a neuronal population whose projections were to be analyzed. The amino acids are taken up, incorporated into proteins, and shipped down the parent axon and all its collaterals to the terminals, where they accumulate. The precise injection site and projection pattern of the labeled neurons can then be reconstructed from autoradiograms of a series of sections cut through the brain. This method had two great advantages: it proved to be much more sensitive than the older lesion methods (it showed many more pathways or circuit elements), and it did not involve fibers-of-passage because axons do not contain protein synthetic machinery. This critical feature eliminated the false positive results so common with lesion methods.

The major disadvantage of the autoradiographic method was that the morphology of labeled projections (axons and terminals) was not observed directly. Instead it had to be inferred from a pattern of silver grains. This problem has since been overcome with the introduction of other purely anterogradely transported tracers, most notably *Phaseolus vulgaris*-leucoagglutinin (PHAL). This protein tracer is detected with an antibody (immunohistochemically), and labeled axons from the very clearly defined injection site (group of neurons generating the labeled projection pattern) are labeled with the clarity of a Golgi impregnation. Thus, the PHAL method amounts to an experimental Golgi method for long projections between neuronal cell groups.

A second general strategy in experimental pathway analysis was initiated by Bernard von Gudden in 1879. He observed that when certain cranial nerves are avulsed near their origin in newborn animals, retrograde degeneration may be observed in the brainstem motoneurons that give rise to the nerve. This demonstrated that, in principle at least, the origin of pathways could be demonstrated by retrograde cell degeneration, just as the course and termination of pathways could be examined by anterograde axonal degeneration. In practice, however, very few pathways in the central nervous system of adult animals undergo obvious retrograde degeneration. If an axon is cut after it generates a minimum number of collaterals, there is typically little obvious retrograde, cell body degeneration—which is referred to as *chromatolysis*.

The solution of this problem also awaited the early 1970s, and this time used fast retrograde intra-axonal transport of injected markers. There are many such markers, including the protein horseradish peroxidase (HRP), and a wide variety of fluorescent dyes. They can be taken up by axon terminals and transported back to the cell bodies of origin, which can be observed by a variety of methods in histological sections under the microscope. This is an exceptionally powerful technique, although virtually all known tracers may be taken up to a greater or lesser extent by fibers-of-passage. This confounds the interpretation of results, but the best solution is to inject anterograde tracers into retrogradely-labeled cell groups to confirm or discount the findings with an independent method. All pathways should eventually be subjected to both anterograde and retrograde tracer analysis because each method reveals different features of the pathway, and the methods confirm one another.

Today, anterograde and retrograde tracer analysis of neural networks is combined in the same sections with histochemical methods to determine neurotransmitter content and other chemical features of particular pathways. Furthermore, these combined methods can also, with a great deal of patience, be applied at the electron microscopic (ultrastructural) level to establish the structural arrangement of synaptic interactions. And appearing on the horizon is a whole new generation of methods for analyzing neural circuits based on genetic engineering. The basic idea here is to take advantage of unique gene expression patterns in particular classes of neurons to generate endogenous tracer molecules restricted to that class.

READINGS FOR APPENDIX C

Cajal, Santiago Ramón y. *Histologie du système nerveux de l'homme et des vertébrés*, vol. 1. Translated by L. Azoulay. Maloine: Paris, 1909. For American translation by N. Swanson and L.W. Swanson, see *Histology of the Nervous System of Man and Vertebrates*, vol. 1 (Oxford University Press: New York, 1995). Chapter 2 has an excellent review of older methods.

Clarke, E., and O'Malley, C.D. *The Human Brain and Spinal Cord: A Historical Study Illustrated by Writings from Antiquity to the Twentieth Century*, second edition. Norman: San Francisco, 1996.

Haymaker, W., and Schiller, F. (eds.) *The Founders of Neurology: One Hundred and Forty-Six Biographical Sketches by Eighty-Eight Authors*, second edition. C.C Thomas: Springfield, 1970.

Nauta, W.J.H., and Ebbeson, S.O.E. (eds.) *Contemporary Research Methods in Neuroanatomy*. Springer-Verlag: New York, 1970.

Rasmussen, A.T. *Some Trends in Neuroanatomy*. Brown: Dubuque, 1947. This is a terrific historical overview of neuroanatomical strategies.

Swanson, L.W. *Brain Maps: Structure of the Rat Brain: A Laboratory Guide with Printed and Electronic Templates for Data, Models and Schematics*, second edition, with double CD-ROM. Elsevier: Amsterdam, 1998–1999.

Swanson, L.W. A history of neuroanatomical mapping. In: A.W. Toga and J.C. Mazziotta (eds.), *Brain Mapping: The Applications*. Academic Press: San Diego, 2000, pp. 77–109.

Glossary

ACTION POTENTIAL The all or none electrical signal that is transmitted along axons; also known as a nerve impulse or spike.

AMACRINE PROCESS As defined originally by Cajal, an extension of a neuron that on functional grounds acts as an axon and a dendrite because it forms reciprocal synapses with another amacrine process; thus, amacrine processes can transmit impulses in either direction through a neural circuit.

AXON The single output process of a neuron that almost always has collaterals arising at essentially right angles; the dendrites of most invertebrate neurons arise from the axon instead of the cell body (as in most vertebrate neurons).

BASAL GANGLIA OR NUCLEI The ventral, nonlaminated subdivision of the vertebrate cerebral hemisphere (telencephalon, endbrain); its two major divisions are the striatum and pallidum.

BOUTON (FRENCH FOR BUTTON) The presynaptic swelling of an axon; a terminal bouton is at the end of a fiber whereas a bouton-of-passage is a short spine-like arrangement along the course of an axon.

BRAINSTEM The adult derivatives of the hindbrain, midbrain, and interbrain vesicles of the embryonic neural tube; the cerebral and cerebellar hemispheres are attached to it dorsally by way of thick fiber tracts called peduncles, and it extends uninterrupted into the neck and then trunk of the body as the spinal cord.

CELL TYPE Like trees, neurons fall into different types or "species" based on their size, shape, and location, although the fundamental criterion for distinguishing neuronal cell types is connections—their outputs and inputs; there are on the order of thousands of neuronal cell types in the vertebrate central nervous system and countless varieties of these basic cell types.

CENTRAL NERVOUS SYSTEM The brain and spinal cord; the cerebrospinal axis.

CENTRAL PATTERN GENERATOR A neural circuit that generates a patterned output from the motor system.

CENTRAL RHYTHM GENERATOR A type of central pattern generator that produces a continuous or an episodic rhythmical output.

CEREBRAL HEMISPHERE The adult derivative of the telencephalic or endbrain vesicle of the embryonic neural tube; it has two basic divisions in mammals—the cortex (which is laminated) and the nuclei or basal ganglia (which are nonlaminated).

CEREBRUM See CEREBRAL HEMISPHERE.

CONVOLUTION See GYRUS.

DENDRITE A neuronal process (thin extension) that conducts electrical impulses toward the axon (compare with amacrine process); they taper gradually and branch at acute angles.

DIENCEPHALON See INTERBRAIN.

DISTAL Away from a reference structure (like the brain, or a neuronal cell body); as opposed to proximal.

ELEMENTARY CIRCUIT OR NETWORK A model based on the minimal number of neurons required to show the essential organization of a circuit or network.

EPHAPSE An electrical synapse that allows ions to flow in either direction between two neurons.

ETHOLOGY The biological study of behavior.

FIBER-OF-PASSAGE An axon passing through a region without forming synapses.

FISSURE See SULCUS.

FOREBRAIN The endbrain (telencephalon or cerebral hemisphere) and interbrain (diencephalon); standard definition, based on embryology.

GANGLION A distinct mass of neurons in the peripheral nervous system; for historical reasons, the term is still attached to a number of cell groups in the central nervous system, although this usage is gradually disappearing.

GLIA The supporting cells of the nervous system, usually divided into astrocytes, oligodendrocytes, and microglia; besides neurons, neural tissue also contains vascular cells (including capillary endothelial cells and mast cells).

GOLGI TYPE I NEURON A projection neuron; that is, a neuron whose axon leaves the parent cell group and courses to other more or less distant cell groups.

GOLGI TYPE II NEURON A local circuit interneuron.

GRAY MATTER A gross anatomical term referring to parts of the central nervous system that are dominated by the presence of neuronal cell bodies; in fresh tissue they appear gray to the naked eye (compare with white matter).

GYRUS A rounded elevation on the surface of the cerebral cortex, accompanied by one or more sulci or indentations; also referred to as a convolution.

HOMEOSTASIS The dynamic state of equilibrium in the body with respect to its various functions and the chemical makeup of its fluids and tissues (for example, body temperature, blood pressure, and blood glucose levels).

HORMONE A molecule that is secreted into the blood to act wherever corresponding receptors are found throughout the body.

INTERBRAIN The thalamus and hypothalamus, which form the rostral end of the brainstem (and the caudal part of the forebrain).

INTERNEURON In the original sense, any neuron intercalated between a sensory neuron and a motoneuron; local circuit interneurons have an axon that ramifies entirely within its parent cell group, whereas projection interneurons have an axon that extends outside the parent cell group to more or less distant regions.

ISOCORTEX A vast region of cerebral cortex that has six layers, or at least passes through a six-layered stage during embryogenesis; see NEOCORTEX.

LATERAL Away from the midline, as opposed to "medial."

LOCAL CIRCUIT (INTER)NEURON An interneuron whose axonal ramifications stay entirely within the parent cell group.

MEDIAL Toward the midline, as opposed to "lateral."

METAMERE See SEGMENT.

NEOCORTEX A term introduced around the beginning of the twentieth century based on erroneous evolutionary theories and inadequate data; see ISOCORTEX.

NERVE NET A diffusely distributed arrangement of neurons that are typically interconnected via amacrine processes; the first nervous system to evolve was a nerve net, and examples have survived in the human nervous system (for example, in the retina, olfactory bulb, and gut).

NEURAXIS Central nervous system or cerebrospinal axis.

NEURON A nerve cell; the fundamental unit of neural circuits or networks that establishes functional contacts with other cells by way of chemical or electrical synapses.

NEUROPIL A region of neural tissue that consists mostly of cellular processes rather than cell bodies; usually characterized by abundant synapses.

NEUROTRANSMITTER A chemical messenger that is released from an axon terminal (synaptic ending) and then diffuses through the extracellular fluid to produce a response in a postsynaptic cell(s), usually by interacting with a specific receptor.

NUCLEUS (1) a nonlaminated cell group in the central nervous system (term introduced in 1809 by Johann Reil, who preferred it to "ganglion"); (2) the large organelle in the cell body that contains the chromosomes (term introduced in 1833 by Robert Brown).

PERIKARYON The part of the neuronal cell body (soma) that surrounds the nucleus.

PROCESS A thin extension from a cell body; for neurons they are axons, dendrites, or amacrine processes.

PROXIMAL Toward a reference structure (like the brain, or a neuronal cell body); as opposed to "distal."

SEGMENT In topographic or regional anatomy, a modular unit that is repeated serially down the longitudinal axis of the body during early embryogenesis; toward the later stages of development these homologous units, which are also called "metameres," may undergo secondary modifications so that in the adult all segments are not necessarily identical.

SOMA The cell body of a neuron, including the nucleus and surrounding cytoplasm, which is bounded by the plasma (outer) membrane (plural: somata).

SULCUS A groove or furrow on the surface of the brain; a deep sulcus is usually referred to as a fissure.

SYNAPSE As defined by Charles Sherrington (1897), a functional contact between a neuron and another cell; chemical synapses are by far the most common (see NEUROTRANSMITTER), but electrical synapses (ephapses) are also found, especially during embryogenesis.

TELENCEPHALON See CEREBRAL HEMISPHERE.

TELEOLOGY Explains the past and present in terms of the future, in contrast to mechanism, which explains the present and future in terms of the past; teleology is associated with vitalism in biology; it is the theory of purposes, ends, goals, final causes, and values—the Good (E.S. Russell, 1916).

TERMINAL Axon terminal; the swelling at the end of an axon or axon collateral that forms the presynaptic element.

VARICOSITY A swelling along an axon that may or may not form a synapse; some dendrites are also varicose.

WHITE MATTER A gross anatomical term referring to parts of the central nervous system that are dominated by fiber tracts; in fresh tissue they appear white to the naked eye because many axons are surrounded by a whitish myelin sheath that is mostly lipid (compare with GRAY MATTER).

Index